SAFETY IN BIOLOGICAL LABORATORIES

Edited by
C. H. Collins

A Wiley–Interscience Publication

Published on behalf of the
Institute of Biology
20 Queensberry Place
London SW7 2DZ

JOHN WILEY & SONS
Chichester · New York · Brisbane · Toronto · Singapore

Copyright © 1985 by John Wiley & Sons Ltd.

All rights reserved.

No part of this book may be reproduced by any means, or transmitted, or translated into a machine language without the written permission of the publisher.

Library of Congress Cataloging in Publication Data:

Main entry under title:

Safety in biological laboratories.

'A Wiley–Interscience publication published on behalf of the Institute of Biology.'
Includes index.
1. Biological laboratories—Safety measures.
I. Collins, C. H. (Christopher Herbert) II. Institute of Biology.
   QH323.2.S24   1985   574'.028'9   85-16908
ISBN 0 471 90833 9 (pbk.)

British Library Cataloguing in Publication Data:

Safety in biological laboratories
  1. Biological laboratories—Safety measures
  I. Collins, C. H.
  574'.028   QH323.2

  ISBN 0 471 90833 9

Printed and bound in Great Britain

This book is due for return not later than the last date stamped below, unless recalled sooner.

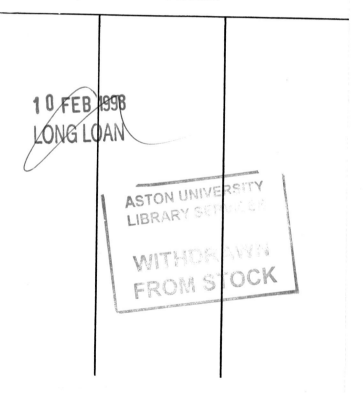

Contents

Preface vii

1. The law and biological laboratories 1
   M. E. Cooper
2. Animals in the classroom and laboratory 11
   P. N. O'Donoghue
3. Supply, collection, and use of living organisms for teaching purposes in schools 25
   J. D. Wray
4. Free-living animals hazardous to man 31
   C. L. Meredith
5. Hazardous plants 37
   C. L. Meredith
6. Safety in microbiology 43
   C. H. Collins
7. Studies in the rural environment 59
   R. C. Clinch and T. E. Tomlinson
8. Planning biological laboratories 67
   K. Everett
9. Safety in the use of chemicals 75
   F. Grover

Index 93

Preface

The Health and Safety at Work etc. Act 1974 applies to virtually all teaching and research laboratories. A number of guidelines and codes of practice have been produced to help people working in laboratories but these are either closely related to particular specialisms or of too general a nature to satisfy professional needs. 'Biology' embraces a diverse and growing group of subjects and it is not surprising that a wide range of threats to safety may sometimes be offered to its students and practitioners. Biologists often need advice on problems, perhaps bridging several biological disciplines, that are not covered in existing publications or are not treated at an appropriate and helpful level.

The Institute of Biology has already contributed to the establishment of safe, sensible working practices by its publication *Safety in Biology Field Work—Guidance Notes for Codes of Practice* (1980, 2nd edition 1984). The success of this publication has now led the Institute to extend its safety advisory activity into laboratories and classrooms.

The underlying philosophy is to reconcile the need to codify and guard against hazards—to achieve safe practice—but at the same time to avoid undue restriction of the teaching and practice of biology. The study of the life sciences, and the educational processes and practical technologies stemming from them, should be made safe without being stifled in the process.

Safety in Biological Laboratories offers advice on a variety of aspects of safe working in biology, on laboratory design and equipment, and on the legal obligations of professional biologists. It makes no pretence at being comprehensive but it presents advice in a form which is immediately helpful, realistic and sensible. It will also serve as the foundation of a continuing contribution by the Institute to safety in biology and its applications.

J. L. Harley
President
Institute of Biology

The views expressed in this book are those of the authors and do not necessarily represent those of their employers.

1 The law and biological laboratories*

M. E. Cooper

THE HEALTH AND SAFETY AT WORK ETC. ACT 1974

The Health and Safety at Work etc. Act 1974 imposed new responsibilities for health and safety and welfare in biological laboratories.

Until this Act came into force on 1 April 1975 the law was concerned largely with redress for harm caused deliberately or negligently and there existed specific safety legislation only in areas such as offices, shops and mines and in respect of certain dangerous materials. Individual bodies may have voluntarily produced safety rules but there were relatively few legal obligations which affected biological laboratories.

The Health and Safety Act now requires those responsible for biological laboratories to formulate and implement codes of safe practice, to be generally well informed upon all aspects of current law and practice relating to safety and to conform to standards set by the Health and Safety Executive. A new body of specialists is emerging—those who make safety a full- or part-time occupation.

APPLICATION OF THE ACT TO BIOLOGICAL LABORATORIES

The Health and Safety at Work Act applies overall general requirements regarding health and safety to all forms of employment (except domestic service in private homes) and therefore applies to biological laboratories

---

* This chapter discusses briefly the law most relevant to these guidelines. It should not be relied upon alone as a source for specific legal information but only in conjunction with the legislation, specialist literature, and professional advice.

at all levels of education as well as in research and industry. The Act affects all persons at work—employers, employees, the self-employed and even those who, although not in employment, may be affected by activities at work.

The duties imposed by the Act may be summarized as follows:

Employers All employers must ensure the health, safety, and welfare of their employees. This applies not only to their work and training but also to premises and environment, transport equipment and materials.

Employees All employees must take reasonable care for the health, safety, and welfare of themselves, fellow employees and anyone else likely to be affected by their work. They must cooperate in the fulfilment of the Act's requirements and must not interfere with anything provided for their health, safety and welfare at work.

Those not employed Employers must also take care for the health and safety of others who may be associated with their work activities. People falling into this category and who may be associated with a biological laboratory include pupils, students, research workers supported by a grant, visiting workers or teachers, social or business visitors, contractors' employees.

In addition, the duty extends to members of the public who may be affected by an employer's activities, for example, by the escape of pathogenic organisms.

Persons in control of premises made available to persons who are not their employees as a place of work where they may use plant or substances provided have responsibility for the safety of the premises and any plant or substance provided.

Self-employed persons Each self-employed person has an underlying duty to carry out his work in such a way as to ensure that he and other persons who may be affected thereby are not exposed to risks to their health or safety. Where a self-employed person works in a laboratory this requirement implies that he must inform other persons in that laboratory of such risks and that he should follow any safety rules or working practices prevailing in that laboratory. For work defined by the Genetic Manipulations Regulations any 'non-employed' person is deemed to be a self-employed person.

It should be noted that these duties are imposed 'so far as is reasonably practicable' and that they do not set absolute standards but those which take into consideration both the degree of hazard involved in a situation and the practicability and expense of its prevention (Cooper, 1981).

IMPLEMENTATION OF THE ACT

An employer who has five or more employees must produce a written statement of his policy, organization, and arrangements for the health and safety of his employees. This may be contained in a single comprehensive document or a collection of rules and codes of practice relating to various premises, types of work and special hazards and produced as booklets, sheets of paper or notices. They must be kept up to date and must cover every aspect of health and safety relevant to any particular employee including not only obvious hazards but also ancillary matters such as personal hygiene and rest areas.

They must be brought to the attention of the employee; this is most effectively done by supplying him with personal copies of the data relevant to his work and requiring a signed acknowledgement of his having received and read them. Alternatively, a general statement may be issued to each employee which refers him to detailed information available in notices.

In the context of a biological laboratory where it is likely that some users are not employees (but, for example, are supported by an independent grant), the authority responsible for it (although not having an express duty to supply a policy document) will best perform its general duty under the Act by issuing users with safety instructions comparable with those given to employees.

An employer must not only formulate but must also implement his safety rules and ensure that health and safety procedures are enforced. This will involve safety committees and safety officers, inspections, training and practices for emergencies. In order to assist an employer in this burden, part- or whole-time safety officers are often appointed from among employees and they are responsible for specific areas of safety, such as a given laboratory or hazards arising from radiation, chemicals or micro-organisms. In addition, under the Safety Representatives and Safety Committee Regulations 1977, recognized trade unions may appoint safety representatives who have a right to be consulted.

Within the bounds of reasonable practicability all hazards should be anticipated. Allowance must be made for personal handicaps and different levels of comprehension in those using laboratories. Health and safety practices should also take into consideration regulations produced under the Act, other codes of practice issued, recognized or used by the Health and Safety Executive, current practices in comparable establishments and other legal obligations.

ENFORCEMENT AND LIABILITY

The Health and Safety at Work Act is enforced by the Health and Safety Executive. Its inspectors make visits to ensure that the safety provisions in biological laboratories comply with the general requirements of the Act and regulations made under the Act as well as any specific standards which the Executive has set. An inspector who is dissatisfied may:

(a) Issue an improvement notice requiring a contravention of part of the safety legislation to be remedied;
(b) Issue a prohibition notice to stop some activity which is causing a risk of serious personal injury until the requirements of the notice have been satisfied;
(c) Prosecute anyone who is in breach of the safety legislation;
(d) Seize from a biological laboratory anything which may be the cause of imminent risk of personal injury and have it destroyed or otherwise rendered harmless.

Liability and prosecution for breach of the safety legislation fall primarily upon the employer, which is often a company or other legal entity, such as a university or local authority. In addition, any of their officers, such as a director, who have direct responsibility for putting the law into effect are open to prosecution. While employees, including safety officers, are rarely prosecuted, the power to do so exists. Safety representatives, however, are exempt from prosecution when acting under the Safety Representatives and Safety Committee Regulations 1976.

LEGISLATION DEALING WITH SPECIFIC ASPECTS OF SAFETY

Other legal aspects of safety which must be observed in the biological laboratory (Cooke, 1976) include:

Fire Prevention
The provisions of the Fire Precautions Act 1971 are gradually being applied to laboratories. Their precautions have to be approved by local fire authorities which may lay down conditions stated in a laboratory's fire certificate, which must thereafter be maintained. The certificate has to be kept on view and the fire authority has power to carry out inspections. Special provisions may be made regarding dangerous substances—for example, the Hazchem Scheme operated by the London Fire Service.

Dangerous Substances
Dangerous micro-organisms The Health and Safety (Dangerous Pathogens) Regulations 1981 restrict and control the use of the organisms (and their derivatives) listed in Schedule 1, such as rabies virus and Lassa fever virus. The regulations have transferred the responsibility for dangerous pathogens from the DHSS to the Health and Safety Executive and came into force on 1 September 1981.

The importation of animal pathogens must be licensed by the Ministry of Agriculture, Fisheries and Food under the Importation of Animal Pathogens Order 1980.

Proposed work involving genetic manipulations must be reported to the Genetic Manipulation Advisory Group of the DHSS and to the Health and Safety Executive in accordance with the Health and Safety (Genetic Manipulation) Regulations 1978.

Radioactive substances The supply, use, storage, transport, and disposal of radioactive substances are subject to the Radioactive Substances Acts 1948 and 1960. Numerous codes of practice have been formulated for this field over the years. New regulations are expected (Taylor, 1981; Rideout, undated). Particular note should be made of the Code of Practice for the Protection of Persons Exposed to Ionising Radiations in Research and Teaching (HSE, 1981).

Medicines and poisons There are substantial restrictions on the acquisition, possession, and use of medicines and poisons. The former are generally restricted to the medical, veterinary, and dental professions unless they are obtained by a prescription. Poisons usually have to be obtained on a signed order. However, both medicines and poisons can be obtained by scientific research and educational bodies, although the former should be administered in accordance with the directions of a doctor, dentist or veterinary surgeon.

Great care must be taken to control the storage and availability of these substances. The production and testing of new medical products must be performed under licence.

Relevant legislation (as amended) includes: the Medicines Act 1968; the Medicines (Veterinary Drugs) or (Products Other than Veterinary Drugs) (Prescription Only) Orders 1983; the Misuse of Drugs Act 1971; the Misuse of Drugs Regulations 1973; the Poisons Act 1972; Poisons Rules 1978; Poisons List Order 1978.

Hazardous chemicals The storage, transport, and use of many chemicals are regulated by the Petroleum (Consolidation) Act 1928 and the Explosives Act 1875-1923. Such substances are also subject to the Packaging and Labelling of Dangerous Substances Regulations 1978. The use of polychlorinated biphenyls at work is restricted by the Control of Pollution (Supply and Use of Injurious Substances) Regulations 1980. Draft regulations on the Control of Substances Hazardous to Health have been produced by HSE.

Carcinogens Certain carcinogenic substances are available only under licence for use in scientific research (Carcinogenic Substances Regulations 1967) (see Chapter 9).

Disposal of waste Various Acts control the disposal of waste. The local authority must be notified of a laboratory's intention to discharge effluent into the public sewer and the consent of the water authority is required for discharge into rivers. The disposal of waste, especially toxic matter, is also controlled by the local authority, for example, the Greater London Council has a Poisonous Wastes Unit. Dumping at sea must be licensed; only limited quantities of radioactive material may be disposed of via normal drainage.

Relevant legislation includes: the Public Health (Drainage of Trade Premises) Act 1937; the Public Health Act 1961; the Control of Pollution Act 1974; the Control of Pollution (Special Waste) Regulations 1980; the Rivers (Prevention of Pollution) Acts 1951, 1961; the Water Resources Act 1963; the Dumping at Sea Act 1974; the Radioactive Substances Act 1960.

Local authorities are also responsible for noise (Control of Pollution

Act 1974) and for smoke (Clean Air Acts 1956 and 1968) and may also have by-laws made under the Public Health Act 1936 to control activities which may adversely affect public health.

Postage Pathological material sent by post must be packed in accordance with Leaflet K681 (inland) or DS061 (overseas) issued by the Post Office.

ANIMALS

Care must be taken for safety in respect of injuries caused by animals and the transmission of disease (see also general legal liability). Certain aspects of the law relating to animals are relevant to health and safety in the biological laboratory.

Dangerous Wild Animals Act 1976 The species of exotic animals listed in this Act may be kept only under local authority licence. The Act lays down general requirements for the animals' welfare and for the safety of their keepers and the general public. No licence is required where such species are kept in zoos, circuses, licensed pet shops and premises registered under the Cruelty to Animals Act 1876. It should be noted, however, that there is no general exemption from licensing for scientific or educational establishments as such.

Importation In the interests of disease control the importation of practically every species of vertebrate is controlled by one or more Orders made under the Animal Health Act 1981. Usually an import licence imposing a quarantine is required.

Zoonoses The occurrence in food-producing species of salmonellosis and brucellosis must be reported to the Ministry of Agriculture, Fisheries and Food, unless this is carried out for scientific or educational purposes and the animals used are disposed of so as to pose no risk to human health (Zoonoses Order 1975).

Plants The importation of plants, plant products and plant pests is controlled by the Plant Health Department of the Ministry of Agriculture, Fisheries and Food under the Import and Export (Plant Health) (Great Britain) Order 1980. The presence of non-indigenous plant pests must be notified and a licence is required to keep or dispose of them or pests which have been subjected to genetic manipulation (Plant Pests (Great Britain) Order 1980). Similar legislation applies to tree pests.

Insects The use and supply of bees is affected by the Bees Act 1980 and the Importation of Bees Order 1980. Other insects may be regulated as pests (see above) or under legislation relating to specific species, such as the Colorado beetle.

There are many potentially dangerous substances for which no specific safety legislation is provided; nevertheless laboratory authorities must, as part of their general duties under the Health and Safety at Work Act and in common law, anticipate such hazards and make provision for their safe use.

GENERAL LEGAL LIABILITY FOR SAFETY IN BIOLOGICAL LABORATORIES

The requirements of the safety legislation discussed so far have been in the nature of criminal law and enforced by prosecution. However, obligations which relate to safety are also to be found in the general principles of the common law; they are further grounds for the provision of sound safety rules in the biological laboratory. Most significant is the law of negligence under which most claims for compensation are likely to be brought, by or on behalf of an individual who is killed, injured or suffers loss in the course of his work, by way of a civil action in the High Court. Nevertheless, a claim might well arise out of the same circumstances for which a prosecution has been brought by the Health and Safety Executive.

A person who is injured because of a failure to take reasonable care of safety may claim compensation for negligence.

An employer is likely to be sued if an accident occurs and he has not provided adequate safety procedures or has failed to enforce them. An employer should therefore aim to avoid claims by ensuring that he has an adequate code for safety and that it is enforced, although ultimately complete protection can only be provided by insurance.

An employee with responsibility for safety, including safety officers, may also be sued if he has failed to follow safety rules or to perform his duties adequately. His employer is more likely to be sued, however, because he will usually be vicariously liable for the employee's negligence (see below), has greater financial assets and has to be insured against liability for negligence (Employers' Liability Compulsory Insurance Act 1969). Trade union safety representatives appointed and acting under the Safety Representatives Regulations are probably exempt from liability (Rideout, undated).

A claim in negligence cannot be maintained for every incident involving harm or loss; the claimant must prove that there was a duty towards him to take reasonable care for his safety, that the harm done could reasonably have been foreseen, that there was a failure in respect of these matters and that as a result he suffered harm. Further, if he had accepted the risk of such harm or had contributed to it in any way, his compensation will be reduced.

An employer has certain other liabilities in common law which may still form a basis for a claim for compensation.

Breach of Statutory Duty

Harm may be suffered by a person as a result of circumstances which also constitute a breach of legislation, such as regulations made under the Health and Safety at Work Act, which imposes a duty. The injured person may base his claim on the breach of duty rather than negligence.

Employer's Liability

There is a common law duty to ensure the general safety of an employee's work.

Vicarious Liability

An employer is liable for the negligence of his employees occurring in the course of their work.

Safety in Buildings

Under the Occupiers' Liability Act 1957 the owner or occupier is responsible for the safety of his premises and structures and must ensure that no one lawfully present on premises is harmed by structural defects.

Injuries Caused by Animals

As an alternative to a claim for negligence, a person injured by an animal may, under the Animals Act 1971, sue its owner or keeper for harm which it has caused, although, if it is a domesticated species the owner or keeper must be proved to have known that it suffered from a propensity, not normally inherent in such species, to cause that sort of harm; if it is not a domestic species he is liable for any harm which it has caused.

CONCLUSION

The health and safety legislation has made it imperative that biological laboratory authorities pay close attention to their obligations in the field of safety. To make proper provision under the Act will also help to provide some protection against accidents which give rise to a common law claim for compensation. Inherent in the legal obligations is the need to keep up to date with current legal and practical information and to review existing safety provisions regularly.

It has been the intention of this chapter to provide an outline of the legal requirements which prompted the production of the guidelines. It is hoped that it will lead to an appreciation of the heavy responsibilities for safety which are imposed by the law and provide the rationale for the practical approach to safety to be found in the ensuing chapters.

REFERENCES AND FURTHER READING

Cooke, A. J. D. (1976). *A Guide to Laboratory Law.* London: Butterworths.

Cooper, M. E. (1981). Legal requirements. In: *Safety in the Animal House.* J. H. Seamer and M. Wood (eds). *Laboratory Animal Handbooks 5*, 2nd edn. London: Laboratory Animals.

Department of Health and Social Security (1976). *Control of Laboratory Use in the United Kingdom of Pathogens Very Dangerous to Humans.* London: DHSS.

Department of Health and Social Security (1979). *A Code of Practice for the Prevention of Infection in Clinical Laboratories and Post-Mortem Rooms.* London: HMSO. (Subsequent bulletins have been issued to this code.)

Health and Safety Executive (1981). *Guidance Notes for the Protection of Persons Exposed to Ionising Radiation in Research and Teaching,* 2nd edn. London: HMSO.

Rideout, R. W. (undated). *Safety in Universities. General Principles of the Law Relating to Safety at Work.* London: Committee of Vice Chancellors and Principals.

Taylor, D. M. (1981). Radiation hazards. In: *Safety in the Animal House.* J. H. Seamer and M. Wood (eds). *Laboratory Animal Handbooks 5*, 2nd edn. London: Laboratory Animals.

2 Animals in the classroom and laboratory

P. N. O'Donoghue

This chapter deals mainly with hazards offered by mammals. Other vertebrates in general pose similar but lesser risks. Invertebrate metazoa tend to come packed, wet or dry, in discrete vivaria and so pose little hazard unless you get in with them or they out with you. The suggestions for safe practice are general and of wide application. While having much in common with the treatment by D. W. Jolly (Hartree and Booth, 1977), this chapter differs in considering a wider spectrum of animal use and in seeing the hazards offered by the animals themselves as only a part of the picture; there are also differences in emphasis. Because the suggestions given here are intended to apply widely, particular cases may need additional advice. An indispensable starting point for any consideration of safety in laboratory animal work is *Safety in the Animal House* (Seamer and Wood, 1981). Where monkeys are involved, *Hazards of Handling Simians* (Perkins and O'Donoghue, 1969) is useful and includes a code of practice which has been developed and updated by the Medical Research Council (MRC, 1985). *The Educational Use of Living Organisms* (Kelly and Wray, 1975) is particularly valuable for work in schools. McSheehy (1976) provides an unrivalled source of information about environmental control and the effects on animals, together with relevant safety considerations.

The importance of avoiding problems by attention to animal health and well-being will be emphasized: access to relevant literature can be gained through the Institute of Biology's *Handbook for the Animal Licence Holder*. The section on health and safety (O'Donoghue, 1980) is largely covered here but is followed by useful references, as is the chapter by Festing, itself a valuable source of guidance on the characteristics and health status of commonly available species. The *Handbook* ends with a glossary and some addresses for further enquiry. The journal *Laboratory Animals* is useful for its scientific and technical articles, for the reference lists that customarily accompany them, and for its book reviews. Most major suppliers of equipment and supplies will advertise in such a periodical. The legislation proposed to replace the Cruelty to Animals Act 1876 may well engender codes of practice, but how far they may go, beyond ensuring observance of the new law, remains to be seen.

Hazards can arise from the animals themselves or from the buildings and equipment used with them. The main danger is mechanical from teeth, horns, claws, hooves, spines or stings, or from poorly designed or poorly maintained premises or apparatus. While microbiological risks such as rabies, Weil's disease and tetanus do exist, the great majority of injuries are without apparent infective complication. However, the important thing about medical statistics—aside from their unreliability—is that no matter how disappearingly small the incidence of a disease, it can represent a personal catastrophe. Accident prevention policy must not be dominated by the more melodramatic possibilities, but they must be taken into account.

Hazards in animal work can never be avoided with absolute certainty but they can be minimized by precautions which are often straightforward and inexpensive. Although considered here under six heads they are interdependent, and the precautionary measures adopted should give a coherent programme of safe practice. Because it may have to be operated by a variety of people—schoolchildren, professors, animal technicians—the essence of the programme must be practical simplicity. All the worry and sophistication must go into its formulation so that safe practice is then achieved without too much reliance on thought or self-discipline. Animals require time and attention every day of the year: it is unrealistic to expect that to be extended by dedication to troublesome routines which may perhaps slightly diminish some already remote hazard. It must be made easy to do the right thing.

These considerations apply as much to pets kept in a classroom as to the most sophisticated laboratory animal house: only the means to achieve safe practice—the distribution of emphasis among the six headings—will vary. But there can never be certainty. The worst injury I have met in an animal house arose when a lighting fitting fell off a ceiling and broke a girl's foot.

It is important that one concerned and informed person (animal curator, director, chief animal technician) be in charge of the area where animals are housed and used, to see that proper practice is adopted. Where animal work is on any scale, especially where complicated procedures or special hazards are involved, a cadre of trained, committed animal technicians is invaluable. They can assure consistent attention to animal well-being, health and temperament, the proper employment of sensible safety aids and routines, and informed comment when the aids and routines proposed are not sensible. By their expertise they can assist in avoiding injury and train scientists and others in safe behaviour. Where involvement with animals is too trivial to justify expert staff, for example in schools, compensatory attention must be paid to the health and behaviour of those animals on acquisition and to what is done with them; some one person must still be in control. Whoever is in charge of an animal area must have the authority to see that safe practice is observed and to exclude anyone who for any reason does not observe these criteria and so creates potential hazards for himself and for others. People who have no direct business there should not be allowed into animal areas.

The six main heads under which safe practice with animals can be considered are as follows:

1. *Housing* The surroundings in which handling takes place, usually a building such as a specialized animal house or room, a classroom or laboratory, but sometimes a pen or paddock.

2. *Caging* The immediate surroundings of the animal, how it is caged or otherwise confined—this can coincide with the building or paddock (for instance with farm animals).

3. *Animal temperament* The nature of the animal itself, whether it has and is likely to use teeth, claws, hooves, horns, venom or bulk, and whether its disposition can be improved by familiarization ('gentling').

4. *Animal health and human health* Any infectious or allergic hazards offered by the animal.

5. *Manipulation* What is to be done with or to the animal.

6. *Personnel* Who is to be in contact with the animal or its products.

Insufficient attention to any one of these is likely to have to be compensated by higher achievement with others. Their relative importance will vary with circumstance, but together they apply to all our contacts with animals, certainly with vertebrates, whether as objects of scientific study, pets or, as in the junior school, something in between. Items 3 and 4, the temperament, physical nature and infective potential, are often the most significant and the simplest to control.

HOUSING

While purpose-built structures are obviously to be preferred, even poor premises can be made safer by comparatively inexpensive attention to access, lighting and internal surfaces. Most important from the safety angle is control of entry of microbes and microbial products that might adversely affect the animals inside and, through them, us. This control is called the 'barrier', a term embracing features of the building, its services, and the work routines. The barrier can be complex and expensive, but the basic concept is a simple one that can and should be applied by all keepers of animals.

THE BARRIER—EXCLUDING PESTS AND CONTAMINATION

Every reasonable effort should be made to exclude creatures that might bring into the animal house transmissible organisms. They may do this in their tissues or gut, on their fur or skin, or mechanically as dirt on feet or beak. The greatest pressure to enter is likely to be exerted by scavengers—rats, mice, some birds—and these pose the greatest risk of carrying infections from other scavengings. Pets from elsewhere, most certainly dogs and cats, should be kept out. Even where they are well-enough controlled not to endanger the contained animals physically, they normally meet a different microbial population and can carry infections transmissible directly or indirectly (for example by soiling food or bedding material) to them and so, perhaps, to man.

It is desirable (but more difficult) to keep out flies and creatures such as cockroaches and crickets, which are more ubiquitous than the daytime inhabitants of buildings suspect. They are mechanical transporters of disease organisms, and cockroaches are known to carry ingested virus safely through formaldehyde fumigation. Such viruses may not usually threaten man directly but could do so indirectly by altering the temperament of laboratory animals that become infected. While a significant hazard might exist in laboratories investigating infections dangerous to man, in educational establishments probably all that is necessary is reasonable housekeeping for the animals and hygiene for ourselves. However, the absence (or relative absence) of scavenging and parasitic arthropods could be taken as an indication of general good practice.

The physical barrier is easier to establish if the building is of brick or concrete, with gaps around pipes and drains carefully (and inedibly) sealed. Wooden walls or floors must be kept in good repair and where necessary reinforced with metal sheet or mesh. Roofs should be birdproof. Rodents can often slip or gnaw their way in under doors, and the usual method of thwarting them is to install a metal sheet about 50 cm high, with the top bent horizontally outwards and then down to baffle leaping animals and avoid the worst accidents to human shins.

Ideally an animal house should have no opening windows. Where they exist they should be screened. Goods entrances can admit unwanted animals. At the least the same precautions should be taken as for other doors, and at best some form of sterilizing or disinfecting lock—dunk-tank, autoclave or steam, ultraviolet light, fumigation or spray chamber—could be provided.

Outside runs are a problem: where possible wire-mesh close enough to exclude rodents and birds should be fitted. Special attention must be paid to any doorway between run and indoor pen, which could breach the barrier of the building itself. Where larger animals are kept in a paddock, little can be done beyond isolating from other members of the same species and if possible excluding foxes and domestic dogs.

Common sense is the watchword. Protect animals from introduced infections and from infestations by fleas and lice and they will not be able to pass them on to you. However desirable from the viewpoint of animal management, more subtle and expensive developments of the barrier—filtered air, sterilized water, irradiated food, and bedding—are, in this country at any rate, unlikely to be of importance to human health. A change of clothes is desirable (see below) but, except where dangerous infections are being studied, showering is more of a social amenity and a reminder of the existence of the barrier than a significant contribution to safety.

GENERAL DESIGN CONSIDERATIONS AND SERVICES

The building itself should offer as few physical hazards as possible and should encourage good behaviour. The more complicated a concept the more ways there are of misunderstanding it, so by and large the building should be kept plain and simple. Where special facilities are required they must be as readily understandable and foolproof as possible.

Good lighting is essential for proper observation of the animals and for safe operation of equipment. Floor cleaning frequently involves a mixture of water and electricity, so that electrical outlets should be of a quality and sited to minimize the hazard. Floors should be self-draining and non-slip. Non-slip floors are more difficult to clean, but the risk of injury by falling far outweighs that of becoming infected. Operating theatres must have ventilation that protects users from anaesthetic agents and sparkproof electrical outlets if flammable vapours might be used. Rooms where other hazardous agents may be employed must be suitable for easy, safe working. Special ventilation, drainage, and room to don and doff protective clothing may be necessary. Where fumigants are to be used, means of releasing them, confining them within the room and clearing them from it must be well planned in advance.

Routes of normal and emergency exit should be unobstructed and clearly indicated, as they may be used by people carrying animals and other bulky or preoccupying loads. For an establishment of any size some means of communicating with people at a distance should be provided. Telephone systems that can be answered without touching, indeed without interrupting work in hand, are excellent.

The building and its equipment should foster calm, deliberate behaviour in the people there, and that will tend to make for tranquil animals. Tense, frightened animals are often more aggressive. No one likes being bitten, and the use of apparatus made of glass or charged with electricity, dangerous gases or fluids becomes more haphazard when one is.

CAGING

Cages and pens should be free from sharp edges or other features on which inhabitants or human visitors can wound themselves. They should not readily degrade in use (or abuse) to generate such hazards. The animals should be securely but comfortably confined, with room to move about — calm, content, unaggressive. The cage should facilitate easy, thorough cleaning. Cage design is of less importance with pets or semipets, where the inhabitant is likely to spend more time outside it and where more time and attention can be expended looking after individual animals.

Cages must allow easy access to animals, preferably from above. Reaching through a little doorway, groping about for an animal and then trying to thread out hand and animal together is a good recipe for bites and scratches. These are likely to make further attempts to handle the animal less confident, in turn unsettling it and making it prone to bite and scratch even more.

With large animals such as pigs and cattle an opportunity for flight must be provided, as they are too heavy for a determined attack to be warded off. Even a genial lean by a bull can be dangerous.

Monkeys pose a special problem. The larger ones can inflict severe injury. Even smaller ones, because of their ingenuity in escaping from cages and their speed with hands, feet and teeth, require special care in caging with facilities for safe removal (for instance by immobilization and sedation). The building in which they are housed should confine them securely if they escape from their cages.

ANIMAL TEMPERAMENT AND CHOICE OF SPECIES

The use for scientific purposes of species commonly favoured as pets no doubt sprang from their availability. A bonus has been that these animals had been selected as being either unaggressive by nature or susceptible to 'gentling', to being tamed. The consequence is that the long-established pets—mice, rats, Syrian hamsters, guineapigs, rabbits, cats, dogs—can, if properly treated from infancy, be handled with confidence. Larger domesticated species, farm animals, have similarly been selected for docility, although they are rarely as cosseted, are allowed behavioural liberties in exchange for marketable qualities, and can be dangerous from sheer size however amiable their intention.

For laboratory and educational purposes there is rarely any reason why such docile animals should not be chosen. Time should be allocated for gentling and no animal should be allowed to become unfamiliar with human company. Caging and housing should be selected to suit the nature of the animal, permitting relative freedom to behave in ways that appear to give it satisfaction: the desire for exercise by dogs is an obvious example.

The use of species outside this usual range should be considered very carefully. With ferrets and other imperfectly domesticated species, safe handling is improved by calm, deliberate, interested attention from an early age. The same can be the case with 'wild' animals, although generally speaking adults will never be altogether reliable. Many simian species are great fun as infants, but few are really safe once they have matured. Even with the offspring of 'wild' animals, success in gentling is likely to be limited—their species and strains have not been selected for domestic compatibility and teeth, claws, spines or venom may at any time be employed in ways outside our expectation.

The use of special clothing or devices to protect against injury by animals is often a confession of failure, and perpetuates or even exacerbates handling difficulties that should be overcome by gentling. However, with the less customary species it may be necessary to recognize the limitations of familiarization—to confess failure—and gloves, nets, 'crush' cages, even tranquillizing darts, may have to be employed. All this is a nuisance and there is a temptation to cut corners, even at personal risk. It is a good reason for sticking to the usual currency of domesticated animals.

Any species may become more aggressive if unwell, or in certain sexual states such as oestrus or when pregnant or nursing a litter.

ANIMAL HEALTH AND HUMAN HEALTH

Infective hazards posed to man by animals—zoonoses—are legion, but cases are rare. Disease can be caused by viruses, mycoplasmata, bacteria, fungi, protozoa, helminths, and arthropods. The agents can be transmitted by excremental contamination, contact, aerosol, or injected by bite or scratch. They may produce manifest disease in the animal or be inapparent or trivial until introduced into man. Some diseases, such as tuberculosis, salmonellosis, and probably some viral conditions, can be transmitted by man to other species and through them to other people. Some are easy to detect, some difficult. Some result in discomfort, others in death. They may be readily susceptible to treatment or intractable. Seen against the infrequency of occurrence, this variety poses a considerable medical problem. There are rarely enough cases of any one condition to attract much attention or generate much expertise.

Avoidance of Infection

Clearly the infections carried by animals are best avoided altogether. With the common species of laboratory animal they can be, as stocks are available that are known to be free of disease transmissible to man and an assurance of this standard should be sought from the supplier. A scheme offering such assurance was devised by the Laboratory Animals Centre (MRC, 1974), and has been succeeded by one operated by the Laboratory Animal Breeders' Association.* Except where there are overwhelming reasons otherwise, only such animals should be bought (or accepted as a present). It seems unlikely that overwhelming reasons will apply to animals in schools. Beware free gifts that may bear disease: goodwill, generosity, kindness, love of animals are no justification for infecting children.

Rats, mice, hamsters, guineapigs, and rabbits of suitable standard can all be purchased without difficulty (although probably not locally—you may need to go to a certified breeder). Local veterinarians may not be able to contribute much with these smaller species but should be able to help considerably with dogs and cats, ensuring their health and freedom from zoonoses. Apart from some very unpleasant parasites that can spread from them, it has been suggested that virus infections of cats are implicated in some human conditions and it would be wise to handle cats with discretion, certainly not to thrust them at children, until the suggestion is resolved.

As with temperament, when less familiar species are used the possibility of hazard is much greater and much less known. The 'Marburg disease', infecting (and killing) man from vervet monkeys, and Lassa fever, recently recognized as a widespread zoonosis in African animals, were completely unexpected. There is no reason to suppose that they will be the last unpleasant surprises. Once again it must be emphasized that the use of animals other than the well-known laboratory or farm species should be undertaken only with very good reason and after careful plans have been laid to contain the animal and any disease it may carry. Even with familiar animals, the universality of salmonellas and *Toxoplasma gondii* should not be forgotten.

Most imported mammals (except herbivores and wholly aquatic species) have to be quarantined for 6 months (life for vampire bats) under the Rabies (Importation of Dogs, Cats and Other Mammals) Order (1974, 1977); where they are not purpose-bred and known to be of a reliable standard this period is a convenient time for investigation and treatment. Rabies itself, although a rarity under such circumstances, is worth the most stringent precautions.

Tetanus One disease hazard inherent in all work with and around animals is tetanus. Although a cure is nowadays often possible, and injections

---

* Honorary Secretary, c/o Charles River (UK) Ltd, Manston Road, Margate CT9 4LT.

of antiserum can be given after the wound has been made, it is more comfortable and more sensible to take prophylactic measures (3 injections of tetanus toxoid spread over about 8 months, followed by a booster injection every 4 or 5 years). This is so painless and free from side effects that it should be adopted by all who work or play with animals, including children.

General precautions   Basic hygiene should be observed at all times when working with animals. Hands and mouths should be kept apart — no smoking, sweets or nailbiting. Food and drink should not be stored or consumed where they might be contaminated: this is not a matter of just issuing an instruction but of making it easy or desirable to obey. People working with animals may smell of them and not be acceptable, or fear they will not be acceptable, in general eating areas, so that a special staff-room or generous time to clean up *en route* to a canteen may be necessary. Every facility should be provided for hand washing on journeys in both directions between staff-rooms, lavatories, and animal quarters. A period of animal handling in schools should also be followed by hand washing, and some form of protective clothing (for example aprons) should be considered.

Apart from protective clothing for special risks, clothing should be provided for wearing in place of 'street clothes' in the animal area to minimize transport of organisms and smells both in and out.

Food and bedding material should be obtained only from suppliers and transporters who realize the possibility of contamination and take effective measures to prevent it.

Without necessarily accepting all ideas about the significance of microbial build-up, it seems sensible every now and then to empty and clean thoroughly each animal room, with formaldehyde fumigation if at all possible. In schools this latter is likely to be impracticable and must be compensated for by more stringent selection of animals.

Deliberate infection   There remains the special case where a hazardous infection is itself being studied. Depending on the agent and mode of transmission, minimal precautions will range from exclusion of a vector or donning of simple protective clothing (impervious gloves and overshoes or boots, gown, impervious apron, cap, visor or *appropriate* protective mask) to special rooms, safety cabinets or negative-pressure isolators. Any appropriate protective vaccination and medical monitoring during the course of the work must be employed. Work with such infections should not be allowed until their potential has been fully understood and suitable precautions taken: the person who wants to use the agent is by no means always the best judge of such matters.

Treatment of Wounds

First aid after injury is a problem. When the skin is pierced by animal or apparatus, should the wound be regarded as a medical emergency and rushed for treatment, or as a mechanical nuisance and have a plaster clapped over it? It must be accepted that someone working by himself late in the evening, at the weekend, on Christmas Day—a commonplace with animal experimentation and maintenance—is under strong pressure to finish and get away. Unless the wound is severe or the wounded has been convinced of an infectious hazard, he is unlikely to delay his departure by the period usually necessary to obtain medical help. In such cases it may be better to provide means of cleaning and covering the wound to protect against subsequent contamination, although some regard this as encouraging avoidance of medical assistance and so risking infectious complications. Every encouragement should be given for even trivial personal accidents to be recorded, and the recording system must not become so elaborate that ordinary people cannot be bothered to operate it. Apart from the importance to individuals of establishing when an illness might have arisen from an accident at work, such records can disclose patterns of trouble not readily seen day to day.

What of medical assistance? What is to be done when somebody comes in with an injury contracted from or near an animal? Unless a known infection is present it is hard to see any course but waiting for the development of symptoms on which a diagnosis could be based (but see tetanus, above).

Allergies

Among common laboratory species rats seem to cause most trouble, with cats and rabbits next. Allergies generally manifest by sneezing and wheezing, although skin reactions may be significant. Those with only small contact with the animals in question can often avoid it altogether, put up with the reaction, or rely on antihistamines or some other treatment. However, allergies tend to be progressive and sensitive people may have to abandon work based on contact with certain species. Particular care must be taken with small children, and allergic reactions in a class may compel discard of a species as a pet.

MANIPULATION

This subject has been dealt with to a large extent in considering animal temperament and health status. For safe and rewarding manipulation animals should have been 'gentled' and accustomed to handling for that purpose. Restraint (for example, for injection) should be well informed, definite, and gentle. Dabbing at an animal is likely to make it apprehensive and aggressive or unpredictable, while at the other extreme restraint should not be synonymous with strangulation. The animal must be held so as to present no danger to the operator, and yet be so little stressed as to be wholly unaggressive on release. Green (1979) offers useful advice on and illustrations of proper handling for injection of laboratory and farm species, and Waynforth (1980) for handling rats. There are national codes of practice for work with radioactive substances, and various draft or private codes for work with, for example, carcinogens.

Proper, mutually agreeable handling of animals should be inculcated at all levels of use, including in schools. Even where specific hazards are remote it means that the handler is that much more in charge of the situation and so better able to deal with unexpected events.

Large animals may need special cages, crates or even buildings to immobilize them for particular procedures. Animals large, fierce or venomous enough to cause real damage should never be handled by one person alone, to allow for being caught off guard, or slipping, or any other accident that could impede withdrawal to safety. Special apparatus, from gloves for small rodents or forked sticks for venomous snakes to 'crush' cages for monkeys, is likely to be needed for exotic or otherwise unusual animals (no one should assume that a British wild rat may be manipulated in the same way as a British laboratory rat of the same species). Even with these aids handling should be directed by the same spirit of kind decision recommended for more direct contact; the animal is then likely to accept future manipulation more philosophically, to mutual benefit. Where special routines are required, for example in dealing with unusual hazards, they can with advantage be written down and displayed, but it must be ascertained that those concerned have read and sufficiently understood them.

PERSONNEL

Sufferers from certain human diseases such as tuberculosis can transmit them to laboratory and other animals, and people can also transmit conditions from their pets, for example ringworm and fleas. Such people should be excluded from the animal house. So should anyone incapable of calm, effective, assured behaviour. No one who is not at ease with animals, or who cannot be instructed to make him so, should be allowed to work directly with them—such a person will be a menace to himself and to his colleagues.

ANIMAL PRODUCTS

These are as dangerous and only as dangerous as the animals from which they derive. Animals known to be free from agents hazardous to man should not be able to offer such hazards in their faeces or tissues, but as there is always an element of uncertainty basic hygiene, hand washing or the use of protective clothing, should always be observed. It is in any case wise to avoid ingestion of or over-intimate contact with faeces and urine, and to treat dusts and aerosols from animals with caution. These precautions should be intensified as the health status of the donor animal becomes more obscure, for example with abattoir material or at necropsy of an animal seen to be unwell or found dead. In such cases instruments should not be washed with bare hands (a source of fatal infection with the Marburg agent). Carcasses and unwanted specimens should preferably be burnt. Mincing them into the sewers is regarded as an acceptable alternative, but is at best only safety by dilution.

All parts and products of animals that may harbour zoonotic agents must be burnt. Cages, containers, and instruments used with them must be autoclaved, adequately steamed or *reliably* chemically disinfected (calling for more thought than many people realize). Indeed perhaps at longer intervals and with less urgency, disinfection is a wise precaution even where there is no identified infection. An alternative is to leave equipment in rooms while they are fumigated: the limits of fumigants with organic materials and closely apposed surfaces must be kept in mind.

HEALTH AND SAFETY EXECUTIVE

The Executive's inspectors monitor observation of the Health and Safety at Work etc. Act (1974). Even where they are unfamiliar with laboratory animal houses they may have knowledge of standards and practices that could with advantage be adopted. The Act provides for the imposition of health and safety regulations and for provision of practical guidance. Codes of practice may be issued or, if prepared by others than the Health and Safety Commission, approved. It would seem sensible for animal users to devise or adopt codes suitable for their purposes, or to contribute to such codes through some scientific or other central organization.

REFERENCES

Festing, M. F. W. (1980). The choice of animals for research. In: *Handbook for the Animal Licence Holder.* H. V. Wyatt (ed.). London: Institute of Biology.

Green, C. J. (1979). *Animal Anaesthesia. Laboratory Animal Handbooks 8.* London: Laboratory Animals.

Hartree, E., and Booth, V. (eds) (1977). *Safety in Biological Laboratories.* London: Biochemical Society.

Kelly, P. J., and Wray, J. D. (eds) (1975). *The Educational Use of Living Organisms: a source book.* London: English Universities Press.

McSheehy, T. (ed.) (1976). *Control of the Animal House Environment. Laboratory Animal Handbooks 7.* London: Laboratory Animals.

MRC (1974). *The Accreditation and Recognition Schemes for Suppliers of Laboratory Animals.* Laboratory Animals Centre Manual Series No. 1, 2nd edn. London: Medical Research Council.

MRC (1985). Statement on the management of simians in relation to infectious hazards to staff. London: Medical Research Council.

O'Donoghue, P. N. (1980). Animal experiments, personal responsibility, and the law. In: *Handbook for the Animal Licence Holder.* H. V. Wyatt (ed.). London: Institute of Biology.

Perkins, F. T., and O'Donoghue, P. N. (eds) (1969). *Hazards of Handling Simians. Laboratory Animal Handbooks 4.* London: Laboratory Animals.

Seamer, J. H., and Wood, M. (eds) (1981). *Safety in the Animal House,* 2nd edn. *Laboratory Animal Handbooks 5.* London: Laboratory Animals.

Waynforth, H. B. (1980). *Experimental and Surgical Technique in the Rat.* London: Academic Press.

3 Supply, collection, and use of living organisms for teaching purposes in schools

J. D. Wray

The use of living organisms in the teaching of biology places two burdens of responsibility upon the teacher. The first is that the pupils are not placed in any danger from the organisms; the second, no less important, is that wild species are not endangered by ill-advised collection and experiment (Kelly and Wray, 1975; see also the Conservation of Wild Creatures and Wild Plants Act 1975 and the Wildlife and Countryside Act 1981).

SUPPLY

Living organisms and material of living origin should be obtained, especially by schools, only from reputable dealers, suppliers or breeders. A scheme offering assurance of quality for the common species of laboratory animals was operated by the Medical Research Council's Laboratory Animals Centre. This has been succeeded by one operated by the Laboratory Animal Breeders' Association (see address on page 19). Supplies from other schools, particularly of vertebrate animals, are acceptable only if the health and quality of the stock can be reasonably assured.

Living organisms supplied to and kept in schools should be easy to handle, and normal in appearance. They should be healthy and free from disease, in particular those diseases transmissible between animals and humans (zoonoses). It should be noted that vertebrate animals can be particularly hazardous in this respect. They should not be poisonous, pathogenic or seriously allergenic (Wray, 1974a, 1975; RSPCA, 1985; see also Chapters 2 and 4).

HAZARDOUS ORGANISMS

Animals

Certain animals should not be supplied to or kept in schools because of health hazards. Such animals include all native wild mammals, monkeys, psittacine birds (budgerigars, macaws, parrots, and parakeets), doves and pigeons, poisonous reptiles, tortoises and terrapins, alligators, crawling insects, especially cockroaches and caterpillars which have hairs that can penetrate the skin and are then broken off ('woolly bear' caterpillars) (Wray, 1974a).

Plants

Plants and seeds should be, as far as is known, free from pests and diseases, in order to restrict the spread of infection to other plants. Many seeds, available commercially, are dressed with pesticides or fungicides which may be very poisonous. These seeds must always be handled carefully (Association for Science Education, 1979a). Many wild and cultivated plants have poisonous or irritant parts, such plants also must always be handled carefully (North, 1967; Tampion, 1977; MAFF, 1979; see Chapter 5).

Dead Animals and Material of Living Origin

Dead animals should not be brought into schools unless prepared as a foodstuff and subject to inspection or for dissection or display. It must be clearly recognized that there may be serious risks of disease in the examination of unprepared dead vertebrate animals brought in from the wild and of birds' nests and owl pellets.

Parasites

Living organisms, or their eggs or larval stages, which are parasitic in vertebrate animals and in particular man, must not be brought into schools unless specially preserved for display.

Micro-organisms

Micro-organisms, especially bacteria and fungi, must be known to be non-pathogenic (Department of Education and Science, 1977; Fry, 1977;

Association for Science Education, 1981; Safety in Education Bulletin No. 1, November, 1981, see Department of Education and Science, 1978; see also Chapter 6).

Collecting from the Wild

The inherent dangers to the collector posed by the environment must always be considered, as must the possibility that the activities of one individual may place others at risk. These problems are dealt with in *Safety in Biological Fieldwork* (Institute of Biology, 1983).

MAINTENANCE

Cages and Containers

All living organisms must be housed in suitable containers which are made in such a way that they are not a hazard to the organisms contained or to the handler (Wray, 1974b,c). Incompatible species must not be housed together. All cages, containers and water bottles must be cleaned regularly, preferably with an appropriate disinfectant. Clean uncontaminated litter and nesting material must be provided for animals, in particular small mammals (Wray, 1974c).

Precautions must be taken to prevent the escape of animals, especially insects and small mammals. Animal stock, especially small mammals, should be kept in a secure animal room where the entry and escape of small mammals can be prevented (Wray, 1974b). There must be no contact between captive animals, especially birds and small mammals, and wild species which are carriers of transmissible disease (zoonoses).

Environment and Equipment

The environment surrounding living organisms (temperature, light, ventilation, humidity) must be controlled to ensure that they are in the best of health. Well-kept animals are less likely to be aggressive.

Equipment having electric lamps, timing devices, aerators or pumps, heaters or thermostats and connected directly to mains electricity must be properly earthed (Electrical Equipment [Safety] Regulations, 1975; Association for Science Education, 1978; Health and Safety Executive, 1983). All electrical installations, for example for aquaria, must conform to any local regulations.

Suitable protective clothing (including gloves) should be worn when washing apparatus, especially if it might be contaminated with micro-organisms, and when handling wild and untamed vertebrate animals.

School Holidays

Proper provision for all animal and plant stock should be provided in the school during holiday periods (Department of Education and Science, 1978). It is best if animals are not boarded out in domestic premises during holiday periods. They may be exposed to infections which could be passed on to pupils or students. They may also become aggressive because of the change in environment.

Animal Health

The stock of living organisms should be regularly inspected for signs of ill-health. If serious disease is suspected it is essential, particularly with vertebrate animals, that appropriate professional advice be sought as soon as possible.

Killing and Disposal

Animals must be humanely killed; it is not generally advisable to let pupils witness this (Wray, 1974a,c). Dead bodies, unwanted cultures and other rubbish, for example soiled nesting material and litter, must be disposed of carefully to avoid health hazards. Wherever possible disposal should be by incineration, cultures of micro-organisms being autoclaved first.

USE OF LIVING ORGANISMS IN CLASSROOMS

Teachers must always consider the needs and reactions of the living organisms and the range of sensibility of their pupils and their well-being (Wray, 1974a). This assists in the development of safe practices and correct attitudes in handling living things.

Cultures of micro-organisms, particularly bacteria and fungi, require special handling; they must be kept effectively closed when they are to be examined by pupils (Association for Science Education, 1981a; Fry, 1977). Cultures of bacteria and fungi must be killed before they are opened for inspection by pupils. Exceptions may be made with older students, for example those in the sixth form (see Chapter 6).

EXPERIMENTS INVOLVING PUPILS

Teachers are reminded of their unique legal responsibilities when they are in charge of pupils (Association for Science Education, 1981b). If pupils are involved in any investigations on themselves then the legal responsibility of the teacher becomes complex. The educational advantages of such involvement must be judged in relation to the possible hazards before proceeding.

There must be no pressure on pupils to take part in any investigations on themselves. Teachers should not involve pupils in any procedures outside the range of normal everyday experience (Department of Education and Science, 1978). As far as is possible teachers should ensure that pupils understand the precautions to be taken and the possible consequences of not taking them.

Certain investigations, involving pupils, which are of a potentially hazardous nature include: blood sampling (Association for Science Education, 1979b); tasting chemicals, for example, phenyl thiourea (Department of Education and Science, 1978); tasting foodstuffs; ventilation of the lungs and the use of spirometers; electrical stimulation to produce a sensation of taste (Wray, 1974a); pulsed stimulation (Association for Science Education, 1981b); measurement of blood pressure; physical exercise and the use of ergometers. These investigations should always be conducted with extreme care.

Author's Note: The views expressed in this chapter are those of the author.

REFERENCES

Association for Science Education (1978). *Electrical Safety for the Users of School Laboratories*. Hatfield: ASE.

Association for Science Education (1979a). Use of pesticides in schools. *Education in Science*, **85**, 26. Hatfield: ASE. See also ASE Topics in Safety.

Association for Science Education (1979b). Human blood sampling. *Education in Science*, **82**, 27 and additional note **84**, 33. Hatfield: ASE. See also ASE Topics in Safety.

Association for Science Education (1981a). Safety in school microbiology. *Education in Science*, **92**, 19–27. Hatfield: ASE. See also ASE Topics in Safety.

Association for Science Education (1981b). *Safeguards in the School Laboratory*, 8th edn. Hatfield: ASE.
Association for Science Education (1982). *Topics in Safety*. Hatfield: ASE.
Conservation of Wild Creatures and Wild Plants Act (1975). London: HMSO.
Department of Education and Science (1977). *The Use of Micro-organisms in Schools*. London: HMSO.
Department of Education and Science (1978). *Safety in Science Laboratories*. DES Safety Series No. 2. 3rd edn. London: HMSO. (Safety in Education Bulletins to be issued once or twice a year. Bulletin No. 1 was issued in November 1981, No. 2 July 1982, and No. 3 June 1984.)
Electrical Equipment (Safety) Regulations (1975). S1 1355. London: HMSO.
Fry, P. J. (1977). *Micro-organisms*. London: Hodder and Stoughton for the Schools Council.
Health and Safety Executive (1983). *Electrical Safety in Schools*. London: HMSO.
Institute of Biology (1983). *Safety in Biological Fieldwork—guidance notes for codes of practice*, 2nd edn (revised). D. Nichols (ed.). London: Institute of Biology.
Kelly, P. J., and Wray, J. D. (eds) (1975). *The Educational Use of Living Organisms: a Source Book*. London: English Universities Press.
Ministry of Agriculture, Fisheries and Food (1979). *British Poisonous Plants*. Bulletin 161, 5th Impression. MAFF/HMSO. London: HMSO.
North, P. (1967). *Poisonous Plants and Fungi*. Poole: Blandford.
Royal Society for the Prevention of Cruelty to Animals (1985). *Animals in Schools*. Horsham: RSPCA.
Tampion, J. (1977). *Dangerous Plants*. Newton Abbot: David & Charles.
Wildlife and Countryside Act (1981). London: HMSO.
Wray, J. D. (1974a). *Recommended Practice for Schools Relating to the Use of Living Organism and Material of Living Origin*. London: English Universities Press for the Schools Council.
Wray, J. D. (1974b). *Animal Accommodation for Schools*. London: English Universities Press for the Schools Council.
Wray, J. D. (1974c). *Small Mammals*. London: English Universities Press for the Schools Council.
Wray, J. D. (1975). Safety and the hazards of using living organisms or material of living origin. *Journal of Biological Education*, **9**, 3 and 140.

4 Free-living animals hazardous to man

C. L. Meredith

Many mammals take defensive action when they are handled or disturbed. This action may result in physical injury to the handler, including skin puncture, abrasions, scratches, and bites. Some animals are venomous and their bites or stings may cause discomfort or illness. Anatomical features may not be specifically designed to be defensive (or offensive) but can cause injuries if the animals, even when dead, are handled carelessly. There is always the possibility that wounds inflicted by animals might become infected. The hair or feathers of some animals may invoke an allergic response in some people.

The number of species indigenous to the United Kingdom that offer these hazards is limited. Some are described below; a few are included which are not indigenous but which may be encountered in this country, for example, in animal collections or as unintentional imports.

ANIMALS CAPABLE OF INFLICTING PHYSICAL INJURY

Invertebrates

Some of these animals present hazards when handled or when accidentally stepped on with bare feet:

Cnidaria	(Corals)	Sharp edged or pointed exoskeletons
Brachiopoda	(Lamp shells)	Sharp edged shells
Mollusca	(Oysters, etc.)	Sharp edged and/or spiny shells
Polychaetae	(Marine worms)	Sharp spines
Arthropoda	(Crabs, etc.)	Sharp and/or spiny shells
Echinodermata	(Sea urchins and starfish)	Sharp spines

Annelida The large marine polychaete worm, the king ragworm (*Nereis virens*) can give painful bites. The mouthparts of some indigenous leeches are capable of penetrating human skin. The medicinal leech (*Hirudo medicinalis*) is the only British species which feeds on human blood but the horse leech (*Haemopsis sanguisuga*) and the marine leech (*Pontobdella muricata*) may bite hands put into water containing them. If they attach themselves to skin they should not be pulled off. The application of common salt will detach them.

Mollusca The sharp hinged shells of some of the larger bivalves can pinch when closing. The larger cephalopods such as cuttle fish (*Sepia officinalis*), squid (*Loligo forbesi*), and octopus (*Eldone cirrhosa* and *Octopus vulgaris*) have horny mouthparts and/or horny rims on their suckers which may damage the skin.

Crustacea Some small crustaceans can give quite a painful pinch and larger species, for example, the common lobster (*Homarus gammerus*) and edible crab (*Cancer pagurus*) can be quite dangerous.

A number of animals have pointed structures which may break off and remain embedded in the flesh. If these are not removed they may become infected. An example is the woolly bear caterpillar. Lacerations and abrasions may also become infected. Those contaminated by sea water nearly always become septic if not treated with a suitable disinfectant.

Vertebrates

Some injuries caused by teeth, claws or horns of vertebrates are considered in Chapter 2. Some reptiles and fish can cause injuries.

Reptiles Large lizards and grass snakes can inflict nasty bites when handled. Their teeth are often shed in the flesh and the bites commonly become infected. Treatment of bites requires removal of teeth fragments and disinfection.

Fish The teeth of some fish are also liable to break off and remain in the flesh. The skin of some members of the shark family can abrade the skin if the animals are handled carelessly. Many other fish, including the spiny dogfish (*Squalus acanthius*), the thorn back ray (*Raja clavata*), the skate (*R. batis*), and the eagle ray (*Miliobatis aquila*), have sharp spines on their tail, body, head or fins. Some fish have opercular spines. All of these can cause serious lacerations if handled carelessly whether the fish is living or dead. In addition the electric ray (*Torpedo marmerota*) can give powerful shocks when touched. A net should be used to lift live specimens.

VENOMOUS ANIMALS

Some animals secrete toxic chemicals which are used for attack or defence. The degree of danger from them varies considerably. The skin secretions of amphibians may cause allergic reactions to anyone touching the animals with bare hands. While this is not usually serious it may become so if the toxin is transferred to the eyes. Animals which inject toxin into bite wounds or stings can cause greater harm, sometimes death.

The general health of a person bitten or stung has a considerable effect on the severity of the resulting reaction to the animal toxin. The reactions also vary with individuals and on different occasions. Most stings from venomous animals in the UK cause only local swelling, discoloration of the surrounding skin, and pain or itching. These may be often treated by cooling with ice packs or the application of antihistamine cream. If they are causing an unusual degree of discomfort or if more serious symptoms of allergy (for example, respiratory difficulty, skin rashes or weals, faintness, sickness, pallor and sweating), or severe local, axillary or abdominal pain with extensive swelling or discoloration, fever, cramps, paralysis, mental or visual disturbances, abdominal rigidity, diminished touch or temperature sensation, nausea, vomiting, loss of speech, excessive salivation, difficulty in swallowing, urgent medical aid is necessary. Delay in giving the correct treatment can be dangerous.

Urgent medical aid is also necessary for stings or bites in or near the eyes, nose, mouth or throat when severe swelling may cause respiratory difficulty, even asphyxia. This may be partially reduced by giving the patient sips of cold water or ice to suck; severe difficulty in breathing may require surgery.

It is not uncommon for venomous animals, including snakes, spiders, and scorpions, to arrive in this country in consignments of food. If caught they may be presented, unexpectedly, to biologists. In addition, these and other venomous animals such as tropical marine fish may be deliberately imported. There are considerable risks in keeping such animals (particularly if snakes, spiders or scorpions escape from their containers). Anyone contemplating doing so, either as pets (not to be recommended), or for scientific purposes should be fully conversant with the hazards involved and the procedures to be adopted if bitten or stung.

Invertebrates

Cnidaria Among indigenous species, the beadlet anemone (*Actinia equina*), the opulet (*Anemone sulcata*) and *Sagaria elegans* may cause skin reactions in sensitive persons. The common varieties of jellyfish (*Pelagia noctiluca, Cyanea lamarckii*) can inflict painful stings. Stings from *Cyanea capillata*, sometimes found inshore, can be serious. Local application of household ammonia solution will alleviate the pain from jellyfish stings.

The Portuguese man-o-war (*Physalia physalis*), sometimes brought to British waters by wind and tides, can inflict very dangerous stings. It is often cast ashore, when the stinging tentacles are broken away from the conspicuous float and present a less obvious hazard. Nets should be used to lift live specimens, and skin contact with jellyfish prevented by protective clothing. The stinging tentacles can be very long and some are too fine to be seen.

Arthropoda Some British spiders, including common garden spiders (*Araneus diadematus, A. umbricatus*), house spiders (*Scotophaeus blackwalli, Streatoda bipunctata, Teganaria* spp.), *Clubiona carticalis, Dysdera erythina, D. crocata, Amaurobius similis* and probably Lycosidae (wolf spiders) are capable of biting through human skin. Reactions vary from a 'pin prick' sensation to swelling, irritation, and loss of sensation lasting for several weeks. Antihistamine or calamine creams will help reduce the pain.

Spiders move very quickly and are difficult to handle safely while active. Quite often spiders are accidentally imported with food or freight from overseas. It is unwise to touch any large or unusual species of spider found in the vicinity of airports or docks.

Several species of Apoidea (bees), Vespoidea (wasps and hornets) and, if held, Ichneumonoidea (parasitic wasps), can inflict painful stings. They should be handled with care. When working in the vicinity of hives or nests the appropriate bee keepers' protective clothing should be worn and clothing that is loose enough for insects to crawl into should not be worn. Antihistamine cream or calamine lotion help to alleviate pain, as do the time-honoured remedies of household ammonia or sodium bicarbonate for bee-stings and vinegar or lemon juice for wasp and hornet stings.

Care is also necessary when handling the 'backswimmers' (*Notonecta* spp.) and great diving beetle (*Dytiscus marginalis*); they can inflict painful bites.

Vertebrates

Fish The greater weever (*Trachinus draco*) and lesser weever (*T. vipera*) have venomous spines on the dorsal fins and gill covers which cause exceedingly painful wounds if touched. Usually partly buried in the sand in shallow water, including rockpools, they are easily stepped upon. The scorpion fish (*Scorpaena porcus*), sometimes found around our coasts in similar habitats to weevers, also has spines which can inject a potent venom. The sting ray (*Dasyatis pastinaca*) can inflict extremely painful wounds with its venomous tail spine.

Any wounds caused by fishes should be allowed to bleed freely before they are dressed. Hot water (50–60 °C) will help to relieve the pain. As with wounds caused by marine invertebrates, there is a marked hazard of sepsis.

Amphibians The skin secretions of the common toad (*Bufo bufo*), and of some imported toads, may irritate the nose, mouth, and eyes. The hands should always be washed after handling these toads.

Reptiles The adder or viper (*Vipera berus*) can inflict painful bites which in some cases can cause serious illness or even death. Bites on the face, neck or back are particularly dangerous. Symptoms of systemic poisoning may be delayed and this provides an opportunity for anyone who is bitten to seek urgent medical attention, i.e. anti-venin serum therapy. Skin contact with venom may cause allergic reactions.

It is dangerous for anyone (however expert) to attempt to pick up venomous snakes by hand, even when wearing protective gloves. The correct procedures for so doing are given in the UFAW (1976) handbook.

First aid treatment for snake bite Most people bitten by snakes are very shocked and it is important to reassure them. Only 50 per cent of persons bitten by *Vipera berus* in UK develop any symptoms of systemic poisoning (and less than 1 per cent prove fatal). Nevertheless, transport to hospital or other source of medical aid should be regarded as urgent. Interim procedures while waiting for transport consist of washing surplus venom from around the wound and if a limb is affected it should be immobilized. If possible the injured person should not be allowed to stand or move around.

PARASITIC ARTHROPODS

A number of these feed on human blood and inject anticoagulants into the skin puncture causing pain and irritation. They include: mites (Acari), fleas (Pulicidae), bugs (Cimidae), lice (Pediculidae), ticks (Ixodidae), mosquitoes (Culicidae), midges (Ceratopogonidae), some flies (Tabanidae) and the stable fly (*Stomoxys calcitrans*). The cleg (*Haematopoda pluvialis*) flies silently and gives painful bites and the sheep ked (*Melophagus ovinus*) will also bite man.

Ticks must never be pulled off the body, as the mouthparts will break off in the flesh. The application of ether, chloroform, or of glue or nail varnish containing an acetate solvent will make the tick drop off.

An additional hazard from partially parasitic animals which feed on human blood is the transmission of diseases but in the UK the incidence of vector-borne disease is now negligible.

ALLERGIES

The hair, feathers, frass, and secretions of a number of animals can cause allergic reactions to sensitized persons. Sensitization can be acquired gradually or occur suddenly either on initial exposure or after repeated exposure to the specific allergen. The hairs from the bodies of spiders or caterpillars are a frequent cause of allergy. Sensitized persons must avoid contact with, or in some cases proximity to, the animals responsible.

OTHER EXOTIC SPECIES

For details of these the reader is referred to the book *Precautions against Biological Hazards* (Imperial College of Science and Technology, 1975).

For References and Further Reading, see page 42.

5 Hazardous plants

C. L. Meredith

A great many plants that are found in the wild, or are cultivated for scientific, teaching or ornamental purposes are poisonous to man if they are eaten, or are capable of causing skin irritation. Others have barbs or spines that can puncture the skin, sometimes painfully, sometimes not, thus permitting the entry of bacteria which may be opportunist pathogens. While most professional botanists will be aware of these hazards, it is clearly the reponsibility of other biologists leading expeditions or forays, or who grow plants in teaching laboratories to ascertain the nature of any hazards and to take steps to protect anyone in their care.

This chapter is concerned with the hazards that may arise from handling plants, as people who work in biology laboratories are unlikely to eat them.

IRRITANT PLANTS

Many plants, either whole or in part, contain substances which are harmful to the skin. Some such as the stinging nettle (*Urtica dioica*) and the small nettle (*Urtica urens*), are covered with conspicuous stinging hairs which penetrate the skin on contact and inject a poison which causes blistering of the skin coupled with intense burning and itching.

Others, such as the pasque flower (*Anemone pulsatilla*), greater celandine (*Chelidonium majus*), and the blister plant (*Ranunculus acris*) contain an irritant juice which, if the leaves are bruised, may contact the skin and cause similar blistering and irritation. The wild parsnip (*Pastinaca sativa*) causes phytophotodermatitis if it comes into contact either directly or indirectly with skin which is moistened by sweat or water. Subsequent exposure to sunlight results in redness of the skin, burning, swelling and the formation of blisters which may be large. The intense redness may persist for months. This condition is frequently misdiagnosed and the correct treatment, which is that necessary for contact dermatitis, not given.

The effects of contact with irritant plant material vary considerably from person to person and can also vary according to season; some plants are not irritant at certain times of the year. One person may experience only slight discomfort while others may develop dermatitis or suffer serious allergic reactions.

Plants in the family Anacardiaceae, for example poison ivy (*Toxicodendron radicans*) and poison oak (*T. quercifolium*), are the most effective in causing skin irritation; they are not indigenous to the UK but have occasionally been imported for scientific or decorative purposes.

The species indigenous to the UK or frequently imported which cause temporary skin irritation or dermatitis are predominantly members of four families: Liliaceae, Amaryllidaceae, Primulaceae and Euphorbiaceae.

Liliaceae and Amaryllidaceae The sap from daffodils, jonquils and other narcissi, and from hyacinths and tulips, is irritant and can cause a condition known as 'tulip finger'. This occurs as painful lesions localized around and under the nail, although the nail itself is not affected. In some cases an itching type of dermatitis, known as 'lily rash', and accompanied by fissuring may occur.

Primulaceae The common cowslip (*Primula veris*) may cause dermatitis in sensitized people. Other species of *Primula*, especially the cultivated *Primula obconica* cause dermatitis in a high proportion of people who come into contact with them. Typically a rash appears which causes intense smarting and itching. The rash may progress to blistering and often spreads from the hands to other parts of the body.

Euphorbiaceae Several of the British spurges (and the imported, decorative 'poinsettia') have an extremely irritant sap (which is claimed by some workers to be carcinogenic). The sap causes severe burning of the mucous membranes and is dangerous if it gets into the eyes.

A list of irritant plants is given in Table 5.1.

Precautions and prevention Many people are immune from irritant plants; those who are sensitive are usually aware of the fact and only they need take full precautions. Protection is obtained by wearing appropriate protective clothing. If it is impractical to wear gloves a barrier cream containing sodium perborate can be applied to the hands. Great care is necessary to avoid irritant sap from getting into the eyes. Eye protection should be worn for procedures that may cause sap to spray out.

Table 5.1 Plants with Saps that can Cause Skin Irritation

Ranunculaceae	
Aconitum napellus	Monkshood
Anemone apennina	Blue wood anemone
A. pulsatilla	Pasque flower
A. ranunculoides	Yellow wood anemone
Clematis vitalba	Travellers' joy
Ranunculus spp.	Buttercups
Cucurbitaceae	
Bryonia dioica	Common bryony
Araliaceae	
Hedera helix	Ivy
Polygonaceae	
Polygonum spp.	
Buxaceae	
Boxus sempervirens	Box tree
Thymelaeaceae	
Daphne mezereum	Mezereon
Euphorbiaceae	
Euphorbia spp.	Spurges
	Poinsettia
Araceae	
Arum maculatum	Cuckoo pint
Dioscoreaceae	
Tamus communis	Black bryony
Papaveraceae	
Chelidonium majus	Greater celandine

The hands should always be washed as soon as possible after handling irritant plants. Any sap which gets on to the skin should be washed off immediately and care taken not to transfer it to the face or other parts of the body.

TREATMENT OF SKIN IRRITATION

Persistent reactions, obvious illness or distress, including pain in the eyes, require immediate medical attention. For minor discomforts, cold water, antihistamine or 'soothing' creams may help, and for stinging nettle rashes, particularly troublesome to young children, those age-old but reasonably efficacious remedies the bruised dock leaf or the juice of an onion should not be discounted.

PHYSICAL DAMAGE

The stems, branches or leaves of some plants, particularly grasses and cereals, have sharp edges which are capable of causing deep lacerations if carelessly handled. The large thorns borne on the woody stems and branches are capable of inflicting very nasty wounds and an added danger with any thorns is the possibility of their breaking and remaining embedded in the flesh.

Heavy falling fruits, particularly if encased in hard and/or spiny capsules, can be dangerous.

PRECAUTIONS

It is advisable to wear trousers and long-sleeved shirts or jackets when moving among tall grasses and in virgin scrubland or forest to guard not only against irritant plants but against sharp and spiny leaves and stems that can cause scratches or even more serious lesions. Gloves are advisable for collecting some specimens. Students should not be encouraged to climb trees to collect fruits or other specimens, nor should they be allowed to throw up sticks, etc., or vigorously shake the branches, in order to dislodge them. Ladders designed for fruit picking, and long-handled pruning shears should be used. Branches of trees and bushes allowed to 'spring back' are a hazard which can be easily eliminated.

Activities that may be acceptable to parents during leisure time ought not to be automatically regarded as safe practices for groups of children or students.

RISKS OF INFECTION

Any wound may become infected but the significance of deep puncture wounds is often overlooked. Infection of such a wound with *Clostridium tetani* may result in tetanus. Treatment is complicated and difficult and recovery is prolonged. It is strongly recommended that all persons who are likely to encounter such hazards, either during their employment or during their leisure, should have a course of anti-tetanus prophylactic injections and should continue to receive injections at such times thereafter as is necessary to retain immunity.

TREATMENT

Most injuries received in this way are susceptible to treatment from the First Aid box, but however insignificant they may appear, all thorns and splinters should be removed. All punctures, cuts, etc., should be cleaned thoroughly and treated with antiseptics. Injuries received from plants (or any other source when working with soil) must never be disregarded.

ALLERGIES

Some plant materials which are small enough to inhale may produce allergies. Examples of this are *Aspergillus* spp. which cause aspergillosis, and spores of *Thermopolyspora polyspora* and *Thermoactinomyces vulgaris* which cause 'farmer's lung'. Individuals may become sensitized by such allergens either rapidly or only after frequent or prolonged exposure to them. An allergic reaction can vary from one individual to another in both intensity and character, and symptoms may include skin rashes and inflammation of the respiratory or digestive tract.

Wind-borne fungal spores, soil algae and pollen grains are allergens and cause sensitized persons to suffer allergic reactions on contact. The concentration of allergens in the atmosphere is a significant factor in the incidence and severity of allergic reactions and for this reason quantities of fungi or flowering plants should not be kept in inadequately ventilated classrooms or laboratories.

In places where the staff are exposed to concentrations of allergens it is essential that they wear special respiratory tract protection, for example, filter masks or respirators.

While a large proportion of the population is able to endure mild symptoms without needing to receive medical treatment, it is nevertheless prudent for individuals to seek medical advice if an allergy is suspected. Respiratory tract irritations left untreated can lead to asthma and other serious conditions. Minor skin rashes, however, may be treated with an antihistamine cream, unless the affected individual is known to have previously suffered an adverse reaction to such treatment. In any case if the rash persists or worsens medical treatment should be obtained without delay.

REFERENCES AND FURTHER READING

Altmann, H. (1980). *Poisonous Plants and Animals.* London: Chatto and Windus.
Behl, P. N., and Captain, R. M. (1979). *Skin-Irritant and Sensitizing Plants Found in India.* Ram Nagar New Delhi: S. Chand & Co. Ltd.
Debelmas, A. M. (1978). *Guide des Plantes dangereuses.* Paris: Pierre Delaveau.
Duchmann, W. B., and Genorde, H. W. (1964). *Symptomology and Therapy of Toxicological Emergencies.* London: Academic Press.
Eshleman, A. (1977). *Poison Plants.* Boston: Houghton Mifflin Co.
Forsyth, A. A. (1968). *British Poisonous Plants.* Bulletin No. 161. Ministry of Agriculture, Fisheries and Food. London: HMSO.
Hardin, J. W., and Arena, J. M. (1974). *Human Poisoning from Native and Cultivated Plants*, 2nd edn. Durham: Duke University Press.
Hunter, D. (1969). *The Diseases of Occupations.* London: English Universities Press.
Imperial College of Science and Technology (1975). *Precautions against Biological Hazards.* London: Imperial College.
Jordan, M. (1976). *A Guide to Wild Plants.* London: Millington Books Ltd.
Kinghorn, A. D. (ed.) (1979). *Toxic Plants.* New York: Columbia University Press.
Mitchell, J. and Rook, A. (1979). *Botanical Dermatology.* Vancouver: Greenglass.
North, P. (1967). *Poisonous Plants and Animals.* London: Chatto and Windus.
Rook, A. (1961). Plant dermatitis—botanical aspects. *Transactions, St John's Hospital Dermatological Society*, **46**, 41–47.
Tampion, J. (1977). *Dangerous Plants.* Newton Abbot: David & Charles.
UFAW (1976). *The UFAW Handbook on the Care and Management of Laboratory Animals* 5th edn. Edinburgh: Churchill Livingstone.
Woods, B. (1962). Irritant plants. *Transactions, St John's Hospital Dermatological Society*, **48**, 75–82.

6 Safety in microbiology
C. H. Collins

The risks of working with micro-organisms in a laboratory vary from none to the possibility of serious illness, depending on the nature and amount of the organisms used and the techniques of handling them.

Until recently micro-organisms were considered to be either harmless, or pathogenic, i.e. capable of causing human and animal disease. Although it is unlikely that some microbes, such as those that fix nitrogen in the soil or reduce sulphates in water, could ever infect man it is now recognized that many of those that were once regarded as harmless are capable of causing disease under certain circumstances.

Micro-organisms may therefore be divided roughly into three groups: (1) probably harmless, (2) opportunist pathogens, and (3) confirmed pathogens. Several categories of pathogens may also be recognized according to the severity of the disease they cause and the ease with which such disease spreads through the community.

ROUTES OF INFECTION

Micro-organisms may enter the body by inhalation, ingestion, inoculation, through the broken or apparently unbroken skin, or through the eye.

Inhalation

Many laboratory procedures result in the breaking of films of fluids containing organisms, and the scattering of tiny droplets, which may be invisible. These are aerosols. Some fall and contaminate hands and benches. Others, the smallest, dry almost immediately and the organisms they contain become droplet nuclei which remain air-borne and are moved about on even quite small air currents. If inhaled they may cause infection.

Ingestion

Organisms may be introduced into the mouth in several ways. Using the mouth to pipette fluids can lead to direct ingestion. Fingers, contaminated by handling spilled cultures or from aerosols, can transfer bacteria to the mouth directly or indirectly, by licking labels, sucking pencils, eating, smoking, nailbiting or handling articles that are subsequently placed in the mouth.

Injection

Apart from the obvious hazard of hypodermic needles, infectious material may be injected by broken culture containers, glass Pasteur pipettes or other broken glass or sharp objects.

Through the Skin

Small abrasions or cuts on the skin which may not be visible to the naked eye may allow microbes to enter the body.

Through the Eye

Splashes of bacterial cultures into the eye may result in local or even severe general infections.

It must be remembered that laboratory infections may arise when an organism enters the body through a route which is not available to it normally. For example, brucellosis, normally acquired by drinking infected milk is, in the laboratory, usually an air-borne disease.

ASSESSMENT OF RISK

Before any general code of practice for work with micro-organisms is proposed it is desirable to assess the risks to the health of the laboratory worker by considering the factors in laboratory-acquired infection. They are:

1. The virulence of the organisms, i.e. their ability to cause infection.

2. The natural or acquired resistance of the host.

3. The numbers of the organism required to initiate infection, which vary from very few to many millions.

4. Effectiveness of immunization and therapeutic measures.
5. The laboratory facilities which constitute the *primary barriers* around the organisms to prevent their escape into the environment.
6. The laboratory workers' expertise and instinct for self-protection which constitute the *secondary barriers* around the person.

All of these should be taken into account in the assessment of risk and therefore in making decisions about precautions that need to be taken in handling micro-organisms. Safety measures designed for serious risks to worker and community would be unnecessary and expensive to apply in circumstances where risks are none or minimal.

CLASSIFICATION OF MICRO-ORGANISMS AND WORK WITH THEM ACCORDING TO LEVELS OF RISK

The first classification of micro-organisms into four classes on the basis of hazard, and which took into account factors 1–4 above, was developed in the United States (United States Public Health Service, 1974). In the United Kingdom the smallpox incidents in 1973 and 1978 (Department of Health and Social Security, 1974, 1980a), and the high incidence among laboratory workers of certain diseases, notably tuberculosis and hepatitis (Reid, 1957; Harrington and Shannon, 1976), contracted from the materials they handled in the course of their work, led to similar classifications (Department of Health and Social Security, 1976, 1978). Meanwhile the World Health Organization (1979, 1983) and the Centers for Disease Control and National Institutes of Health (United States Public Health Service, 1984) published alternative classifications. The system now used in the United Kingdom, produced by the Advisory Committee on Dangerous Pathogens (1984) is a synthesis of all these and is regarded by many practising microbiologists as unnecessarily complicated. (Apart from the United Kingdom and the United States, the rest of the world has adopted the WHO classification.)

Table 6.1 shows the classification of the Advisory Committee on Dangerous Pathogens. The viruses in Hazard Group 4 offer the most risk to laboratory workers and the community and include the haemorrhagic fever viruses. This group contains no bacteria, fungi, protozoa, or helminths. Hazard Group 3 contains those organisms that are most likely to infect laboratory workers by the air-borne route, e.g. tubercle bacilli and brucellas, and those that require only very small doses to initiate infections by other routes, such as the typhoid bacillus. Hazard Group 2 includes those pathogenic micro-organisms that are normally encountered in medical and veterinary diagnostic laboratories, but which are not especially hazardous to skilled and careful workers, while Hazard Group 1 includes organisms that are not usually reckoned to be hazardous at all.

Table 6.1 Categorization of Pathogens According to Hazard

Group 1	An organism that is most unlikely to cause human disease
Group 2	An organism that may cause human disease and which might be a hazard to laboratory workers but is unlikely to spread in the community. Laboratory exposure rarely produces an infection and effective prophylaxis or effective treatment is usually available.
Group 3	An organism that may cause severe human disease and present a serious hazard to laboratory workers. It may present a risk of spread in the community but there is usually effective prophylaxis or treatment available.
Group 4	An organism that causes severe human disease and is a serious hazard to laboratory workers. It may present a high risk of spread in the community and there is usually no effective prophylaxis or treatment.

Advisory Committee on Dangerous Pathogens (1984). Note that this is the official title, but that non-pathogens are included.

There is not space in this publication to list the organisms that the Advisory Committee on Dangerous Pathogens has assigned to Hazard Groups 2–4 (there are no lists for Group 1). The current lists should be consulted.

It must not be assumed, however, that all Hazard Group 1 organisms (i.e. those that are not listed in the higher groups) are harmless and are suitable for use in school microbiology classes. Some such organisms, in high doses, or in people whose immune systems are impaired for some reason, have been known to cause serious disease.

There is also another reason why great care is necessary in the selection of organisms for teaching microbiology in schools. Many microbes that can cause disease are present naturally in the environment, and may also be present in or on individuals without manifesting evidence of infection. A technique that is designed to isolate and demonstrate 'harmless' organisms may also concentrate such pathogens into an infective dose. For this reason culture media and cultural conditions used in school microbiology should ideally be restricted to those that are unlikely to support the growth of such pathogens. Examples are media with a pH of 5 or less, and incubation at room temperature. This cannot always be achieved but it should be possible to avoid enriched media such as blood agar, media that are selective for enteric pathogens, and incubation at 37 °C. Lists of organisms considered to be suitable for school microbiology have been published recently (Association for Science Education, 1981; Department of Education and Science, 1977). A shorter list of organisms suitable for school microbiology is presented in Table 6.2. This list will inevitably be criticized as incomplete. There is no evidence, however, that a large selection of species is necessary to teach the elements of microbiology in the limited time permitted in the curriculum. It is the writer's personal opinion that at the level of school microbiology even this list is excessive and that three organisms only will provide enough experimental work and interest, particularly as they can be related to everyday life. They are *Lactobacillus casei, Saccharomyces cerevisiae* and *Mucor mucedo*.

Table 6.2 Micro-organisms Suitable for Experiments in Schools

Bacteria	Fungi
Acetobacter aceti	*Agaricus bisporus*
Agrobacterium tumefaciens	*Botrytis cinerea*
Lactobacillus casei	*Mucor mucedo*
Lactobacillus bulgaricus	*Penicillium roqueforti*
Photobacterium phosphoreum	*Pythium debaryanum*
Rhizobium leguminosareum	*Rhizopus sexualis*
Rhodopseudomonas palustris	*Saccharomyces cerevisiae*
Rhodospirillum rubrum	*Sporobolomyces* spp.

These micro-organisms may be cultured on media that do not support the growth of pathogens; they grow at room temperature.

PRIMARY BARRIERS: LEVELS OF CONTAINMENT

For each of the categories or risk groups of pathogens outlined above it is necessary to have safety precautions (containment) in terms of accommodation, equipment, methodology and expertise. Hazard Group 4 viruses require a *maximum containment* laboratory, sometimes called a 'special pathogens unit'. Hazard Group 3 micro-organisms are handled in a *containment* (Containment Level 3) laboratory, while work with Hazard Group 2 organisms requires a *basic* (clinical) laboratory. The basic laboratory is, however, too sophisticated—and expensive—for work with the organisms listed in Table 6.2 and it seems necessary to introduce an additional level—the *school science laboratory*. Table 6.3 shows how these four levels of containment relate to the nature of work and the organisms handled.

DESIGN OF LABORATORIES

Only the general principles of the design of microbiological laboratories are outlined here, beginning with the School Science Laboratory. More details are given by Everett in Chapter 8 and in standard publications (Everett and Hughes, 1975; Grover and Wallace, 1979; DHSS, 1980b; Collins, 1983; Lees and Smith, 1984).

School Science Laboratories

These laboratories are suitable only for handling micro-organisms listed in Table 6.2 and those included in the list prepared by the Association for Science Education (Department of Education and Science, 1977).

The laboratory premises should be clean and have a good standard of hygiene. There should be gas, electric power, and running water. Handwashing basins should be available and also paper towels (laboratory sinks should not be used on general grounds of hygiene). A school biology laboratory is more suitable than a chemistry or physics laboratory. Ordinary classrooms are unsuitable as they do not have water or gas.

Basic Laboratories

These laboratories are intended for the handling of micro-organisms in Hazard Groups 1 and 2, which offer low or moderate health risks to workers.

Good and safe design may be achieved only by consultations between laboratory users and design teams. The latter are usually unaware of the nature of the work and of microbiological and chemical hazards. Without consultation, hazards may be enhanced or even introduced.

Table 6.3 Levels of Containment in Relation to Nature of Work and Organisms Handled

Level of Containment	Nature of Work				Organisms in Hazard Group
	School*	Higher Education/ College/QC/RD†	Clinical/Research	Dangerous Pathogens	
School science	×				1
Basic		×	×		1, 2
Containment			×		3
Maximum containment				×	4

* Organisms in Table 6.2.
† Quality control, Research and Development, for example, in food and industrial laboratories.

Walls and floors should be made of easily cleanable material and the area of horizontal surfaces should be kept to a minimum in an effort to reduce the collection of dust. Bench tops should be of laminated plastic and all furniture of sturdy construction. Mechanical ventilation is unnecessary and expensive. No laboratory should be without windows.

Hand basins should be provided in the laboratory; paper towels, not roller towels, should be used. Pegs for protective clothing should be fitted in the laboratory rooms and lockers for outdoor clothing should be outside, in corridors or in other rooms, but no great distance away. A separate room or area for eating, drinking and smoking should be provided.

An autoclave is an essential requirement and should be in the laboratory suite. This, together with refrigerators, incubators, and centrifuges, should conform to the appropriate British Standards, if any, and to government and Health Safety Executive requirements for safe use.

Safety cabinets (see below) may be required but as organisms in Hazard Group 3 are unlikely to be handled and emphasis may be more on product protection (for example, sterility) than on operator protection, Class 2 (laminar flow) cabinets may be more relevant than Class 1 (exhaust protective) cabinets. Such cabinets should conform to BS 5726 (British Standards Institution, 1979).

Containment (Containment Level 3) Laboratories

These are used for handling Hazard Group 3 micro-organisms which may infect laboratory workers by the air-borne route (infectious aerosols) or which are especially hazardous because only small numbers are required to initiate infections by other routes such as accidental injection or ingestion.

The laboratories should be separated from areas to which there is public access, and should be away from normal corridor traffic. The design features should be those of the basic laboratory. Ventilation should be arranged so that there is a constant flow of air from 'clean' areas such as corridors to the containment laboratory, via, if convenient, the basic laboratories.

Safety cabinets, Class 1, exhausting to atmosphere outside the building, and conforming to BS 5726 (British Standards Institution, 1979) are essential. An autoclave should be available in the same building.

Design of these rooms requires consultation with the Health and Safety Executive and the appropriate code of practice (Advisory Committee on Dangerous Pathogens, 1984) should be consulted.

Maximum Containment (Containment Level 3) Laboratories

These are used exclusively for the diagnosis of and research on the most dangerous viruses, those in Hazard Group 4 that are not only extremely hazardous to the laboratory worker but offer considerable risk to the community if they escape.

Essentially these are air-tight containment laboratories physically separated from other areas or buildings, with access through air-locks. A mechanical ventilation system ensures inward air-flows so that escape of air-borne infectious particles is minimized. Exhaust air is passed through a high-efficiency particulate air filter before discharge and all liquid effluents are treated by heat to render them harmless. All manipulations are done in Class 3 safety cabinets (see below). These cabinets connect to one door of a double-door autoclave; the other door opens outside the laboratory. The doors are interlocked with the autoclave mechanism so that both doors cannot be open at the same time, and nothing can pass out from the laboratory without being autoclaved.

These laboratories may be built and operated in the UK only by arrangement with the Health and Safety Executive and the Department of Health and Social Security.

BIOLOGICAL SAFETY CABINETS

Biological safety cabinets are important primary barriers against infection by inhalation. There are three kinds:

Class 1. Open-fronted Exhaust Protective Cabinets

These offer adequate protection to the worker against the inhalation of aerosols containing Hazard Group 3 organisms. These cabinets are usually fitted in hospital microbiological laboratories. The protection offered by them depends on correct installation and maintenance.

Class 2. Vertical Laminar Flow Cabinets of Special Design

These recirculate some filtered air, exhaust some to atmosphere and take in replacement air through the open front. There are various designs. All offer protection from contamination to the material handled and some protection to the worker, depending on design and maintenance. They are mainly used in tissue culture work.

Class 3. Totally Enclosed Exhaust Protective Cabinets

These are gas-tight and fitted with glove ports. They are used for handling Hazard Group 4 viruses when complete isolation of work from worker is required. A high standard of maintenance is essential.

Only cabinets which conform to BS 5726 (British Standards Institution, 1979) should be used. Advice on installation, testing, and maintenance is readily available (British Standards Institution, 1979; Collins, 1983; Clark, 1983).

There are laminar flow cabinets, more correctly called clean air work stations, in which a stream of filtered air passes horizontally from the back to the front of the cabinet over the work-piece and into the operator's face. These are not safety cabinets in the strict sense. They are designed to protect the work from contamination and should not be used for dispensing mammalian tissue cultures which, even if uninoculated, may contain 'slow or inapparent viruses' or even oncogenic agents.

APPROPRIATE EQUIPMENT

Good laboratory equipment, chosen with care by laboratory workers who are prepared to seek expert advice, goes a long way towards containing micro-organisms. Equipment should not produce aerosols (Collins, 1983; Collins and Lyne, 1984).

Centrifuges, stirring, shaking and homogenizing apparatus are particularly prone to this fault. Culture tubes and bottles need to be robust so that they do not break and release their contents if they are dropped. Screw caps release fewer aerosols and splashes when removed from bottles than do press-on or pop-up closures. Plastic Petri dishes are safer than glass dishes. Pipetting devices should be readily available. Plastic Pasteur pipettes are safer than those made of glass. Loops for transferring bacterial cultures should be short, completely closed and of small diameter, as otherwise they will shed their contents. Details of these hazards and methods for countering them have been published (Collins, 1983; Collins and Lyne, 1984).

DISPOSAL OF INFECTED MATERIAL

Care is needed in disposal of infected material. Even if 'harmless' organisms have been used there is always the possibility that pathogens may have contaminated the culture and grown to give an infective dose. It is best to autoclave all cultures and equipment after use (graduated pipettes are an exception, see below) and people who are likely to use autoclaves should be carefully instructed in safe operation. Autoclave manufacturers issue such instructions; these should be displayed near to the instrument. Autoclaves must not be opened after use until the temperature dial registers below 80 °C and even then care is necessary. Visors, obtainable from laboratory suppliers for this purpose, should be worn when autoclaves are opened. Autoclaves need to be tested regularly with thermocouples to ensure that their chambers reach a temperature high enough to sterilize their contents. Biological or chemical indicators may be used on a day-to-day basis, but these are not always entirely reliable as they may deteriorate if not stored correctly.

Disinfectants should be chosen and used with care. Some are harmful to the skin and, if splashed into the eyes, may cause serious damage. Some 'household' disinfectants may not even kill the germs. Hospital-type phenolic disinfectants such as Clearsol, Hycolin, Stericol and Sudol should be diluted according to the manufacturer's instructions. Hypochlorites such as Chloros, Domestos and Diversol Bx should be diluted to make a solution containing 2500 ppm available chlorine.

Disinfectant solutions should be prepared fresh daily or for each occasion. Many dilute disinfectants lose their activity within a few days.

Disinfectants may be used for graduated pipettes, which cannot be autoclaved and, if an autoclave is not available, for cultures of the organisms in Table 6.2. All materials to be disinfected should be completely submerged and air bubbles removed, otherwise microbes will not be killed.

SECONDARY BARRIERS

Protective Clothing

The individual worker may protect himself by wearing suitable overalls or gowns, which should always be fastened (DHSS, 1978). These garments should remain in the laboratory until they are sent for 'hot wash' laundering. They should not be taken home or allowed to come in contact with outdoor clothing. This should not be brought into the laboratory.

These precautions obviously cannot be taken in schools, and if work is restricted to experiments with organisms in Table 6.2 or those listed by the Department of Education and Science (1977) or the Association for Science Education (1981) they are not necessary. It is, however, good training to require or suggest that young persons wear some kind of 'lab coat', apron (or even father's discarded shirt) such as are worn for craft work.

Medical Supervision or Advice

This is normal in clinical and most research laboratories, but is also desirable wherever micro-organisms are cultured. It is, for example, extremely unwise to allow any student who is taking steroids or immuno-suppressive drugs to handle cultures of any micro-organisms, even those commonly regarded as 'harmless'. Advice should also be sought about immunization (DHSS, 1978). In places where pathogens are, or might be, handled it would be wise to consider offering protective inoculations.

General Precautions

Mouth pipetting, even with 'harmless' organisms, should be forbidden, and so should eating, drinking, and smoking in the laboratory. All staff and students should be encouraged to wash their hands before leaving the laboratory. All cuts and abrasions on hands and face should be covered with adhesive plasters.

Training and Supervision

Any person in charge of a microbiological laboratory, at any level, should have received basic training in handling microbes over and above that given to undergraduates. It has become increasingly obvious that good technique is of paramount importance in the prevention of laboratory-acquired infections and that even facilities such as biological safety cabinets will not protect an operator whose technique is of a low standard. Teachers, especially, need to learn techniques and to understand the hazards.

Although safety should be included in training it is frequently omitted or given scant attention. Laboratory workers should be encouraged to attend seminars and courses on the safe handling of micro-organisms.

SPECIAL HAZARDS IN TEACHING MICROBIOLOGY

Supply of Materials

Micro-organisms should be obtained from reliable commercial sources or official culture collections. Unofficial acquisition from 'friends' should be discouraged, as mistakes may occur and pathogens may be given to unsuspecting students. Cultures should not be made from human faeces, from the faeces of cattle, reptiles or of any animal with diarrhoea.

If blood is used for any purpose it should be obtained from a blood bank where it will have been tested for hepatitis.

Risks of Allergy

Care should be taken with cultures of sporulating fungi. Clouds of spores may be released when cultures are open and if these are inhaled by susceptible individuals they may provoke an allergic response.

TESTING DISINFECTANTS

In commercial laboratories an allegedly non-virulent strain of *Salmonella typhi* (NCTC 786) is used for this purpose. The excellent precautions given in the code of practice issued by the British Society for Chemical Specialists (British Association for Chemical Specialities, 1980) for this work should be followed. *Salmonella typhi*, including this special strain, is a Hazard Group 3 pathogen but is not spread by the air-borne route. Special accommodation—a containment laboratory—is needed to handle it, but safety cabinets are not necessary.

The use of *Salmonella typhi* for testing disinfectants is banned in clinical laboratories (DHSS, 1978). In colleges and teaching establishments the techniques for testing disinfectants may be taught or demonstrated by using alternative organisms such as *Pseudomonas aeruginosa* or *Escherichia coli*.

CONCLUSIONS

Only minimal requirements are given above. The reference list provides more detailed information. It must be emphasized that risks of infection in microbiological laboratories may be minimized, and in some establishments avoided altogether by well-planned premises, correct equipment, safe techniques, and personal hygiene. The key elements are, however, good education in methodology and an appreciation of the potential risks of handling microbes.

REFERENCES

Advisory Committee on Dangerous Pathogens (1984). *Categorisation of Pathogens According to Hazard and Categories of Containment.* London: HMSO.

Association for Science Education (1981). Safety in school microbiology. *Education in Science*, **92**, 19-27.
British Association for Chemical Specialities (1980). *Code of practice for the handling of* Salmonella typhi: *NCTC 786 in Industrial Laboratories.* London: BACS.
British Standards Institution (1979). *Specification for Microbiological Safety Cabinets.* BS 5726. London: BSI.
Clark, R. P. (1983). *The Performance, Installation, Testing and Limitations of Microbiological Safety Cabinets.* Northwood: Science Reviews.
Collins, C. H. (1983). *Laboratory-acquired Infections.* London: Butterworths.
Collins, C. H., and Lyne, P. M. (1984). *Microbiological Methods,* 5th edn. London: Butterworths.
Department of Education and Science (1977). *Use of Micro-organisms in Schools.* London: HMSO.
Department of Health and Social Security (1974). *Report of the Committee of Inquiry into the Smallpox Outbreak in London in March and April 1973.* London: HMSO.
Department of Health and Social Security (1976). *Control of Laboratory Use of Pathogens Very Dangerous to Humans.* London: DHSS.
Department of Health and Social Security (1978). *Code of Practice for the Prevention of Infection in Clinical Laboratories and Post-mortem Rooms.* London: HMSO.
Department of Health and Social Security (1980a). *Report of the Investigation into the Causes of the 1978 Birmingham Smallpox Occurrence.* (House of Commons Paper 79-80, No. 668.) London: HMSO.
Department of Health and Social Security (1980b). *Hospital Building Note No. 15: Pathology Departments.* London: DHSS.
Everett, K., and Hughes, D. (1975). *A Guide to Laboratory Design,* 2nd edn. London: Butterworths.
Grover, F., and Wallace, P. 1979). *Laboratory Organisation and Management.* London: Butterworths
Harrington, J. M., and Shannon, H. S. (1976). Incidence of tuberculosis, brucellosis and shigellosis in British medical laboratory workers. *British Medical Journal*, **1**, 759.
Lees, R., and Smith, A. F. (1984). *Design, Construction and Refurbishment of Laboratories.* Chichester: Ellis Horwood.
Reid, D. D. (1957). Incidence of tuberculosis among workers in medical laboratories. *British Medical Journal*, **2**, 10.
United States Public Health Service (1974). *Classification of Etiological Agents on the Basis of Hazard.* Atlanta: Centers for Disease Control.
United States Public Health Service (1984). *Biosafety in Microbiological and Biomedical Laboratories.* Atlanta: Centers for Disease Control.
World Health Organization (1979). Safety measures in microbiology. Minimum standards for laboratory safety. *WHO Weekly Epidemiological Record*, No. 44, 340-342.
World Health Organization (1983). *Laboratory Biosafety Manual.* Geneva: WHO.

FURTHER READING

Anon. (1975). *Precautions Against Biological Hazards*. London: Imperial College of Science and Technology.

Fry, P. J. (1977). *Micro-organisms*. London: Hodder & Stoughton for the Schools Council.

Microbiological Consultative Committee. (Joint Coordinating Committee for the Implementation of Safe Practices in Microbiology) (1982). *Guidelines for Microbiological Safety*, 2nd edn. Reading: Society for General Microbiology.

7 Studies in the rural environment

R. C. Clinch and T. E. Tomlinson

The advice given in this chapter is intended for a wide range of people from students undergoing primary education to those carrying out postgraduate research. It is assumed that the majority will have little or no experience of fieldwork or of the hazards and the law relating to such work. The chapter covers not only the general hazards associated with fieldwork but also the more specific hazards and legislation relating to the use of agricultural implements, pesticides and glasshouse work. Further reading and references can be found in *Safety in Biological Fieldwork* (Institute of Biology, 1980).

FIELD EXPERIMENTS AS EXTENSIONS OF LABORATORY WORK

When transferring experiments into the field it is important to remember that they are being moved from a closed to an open environment, and that there are hazards associated with fieldwork which are not present in the laboratory. It should also be borne in mind that with fieldwork a hazard may be presented to other people who are not directly involved in the work. Some of the general hazards are listed below.

Thunderstorms

Lightning can prove fatal; never stand underneath a tree or close to metal objects which can attract the lightning. If a vehicle is at hand it is advisable to sit inside it and wait for the thunderstorm to pass. Otherwise, stay out in the open; it is better to get wet than be struck by lightning.

Wind

When spraying, wind can cause drift of the spray which can contaminate other crops, or can harm livestock. If the wind is more than a gentle breeze (more than 5 km/h, 3 mph) spraying should not take place.

Lifting

Fieldwork may involve much heavier work than is normally required in the laboratory. It is important to learn to lift correctly by always bending the knees and keeping the back straight, with the weight kept as close to the body as possible so that it almost becomes a part of the lifter when he moves; this is commonly known as 'kinetic handling'. Observance of these few simple rules will be less likely to put a strain on the spine or result in back injury.

Driving

The time of greatest risk is probably when driving to and from the field, especially if that involves a journey on public roads. These journeys can be long and are often undertaken early or late in the day when tiredness or inattention may affect driving. It is essential that as well as driving carefully and observing the Highway Code, special attention should be paid if a vehicle unusual to the driver is being used, or if a vehicle is unusually loaded or is drawing a trailer.

Animals

Both wild and domesticated animals are found in the rural environment. Some can transmit disease (for example, Weil's disease can be transmitted by rats), some are aggressive, and some can be dangerous because of their size. When spraying pesticides it is possible to contaminate animals in the vicinity which could then carry the pesticide elsewhere. Particular care should be taken with dogs which are prone to attack people in protective clothing and dance around in the path of the spray. Animals can also become very agitated and excited by machine noise.

See also the chapters on animals (Chapters 2, 3, and 4).

Power Cables

Cables carrying dangerously high voltages are found both overhead and underground. Adequate clearance must be ensured before moving high loads, and the possibility of buried power cables checked (for instance, with the owner of the land or the local electricity authority) before ploughing, or digging the land, or taking samples with an auger.

Pathogens

A different range of potentially harmful micro-organisms can be found in country areas, leading to (e.g.) 'farmer's lung', which is contracted by exposure to mouldy produce, and brucellosis, which can be caught by drinking the milk from infected cows. Personal hygiene may be no more important than it is in an urban setting, but it may have to be thought out in detail for fieldwork.

Children

Children, with their inquisitive nature and enquiring minds, present a hazard to themselves and care must be exercised when equipment is left unattended. Chemicals must never be left lying around unattended, but must be locked away securely, for instance in the boot of a car. Children present for educational purposes must be carefully supervised.

HORTICULTURAL AND AGRICULTURAL MACHINERY

Tractors

Tractors should only be used by people who have received adequate training on the terrain on which they are expected to drive. Passing a Ministry of Transport driving test is not enough. The correct procedure for starting and stopping the engine should be familiar, including first making sure that the tractor and the power take-off shaft are out of gear. Passengers must never be carried, nor may people be allowed to climb on or off a moving tractor. It is useful to get into the habit of automatically disengaging the power take-off shaft when the tractor is stopped, and never to make adjustments to an implement until the shaft is out of gear.

Overturning Tractors can easily be overturned, especially when fitted with implements which alter the centre of gravity. Other causes of overturning are driving too fast, driving too near to ditches or banks, driving on hillsides or on silage heaps, or winching with a tractor. Prevention is simple: always drive slowly when on rough surfaces; always leave an adequate margin near ditches and banks; and be extremely careful when driving on hillsides. Alteration of the configuration of the tractor can help, for example, adding weight to the front or widening the track of the wheels makes it more stable. Nothing should be hitched to a tractor above the proper drawbar level, and when it is to be driven the clutch should be let in slowly to avoid snatching. If a tractor is to be used for winching, it should be set straight in line with the load and a slow and steady pull employed. Releasing the accelerator will arrest a tendency to tip over.

Cabs Under the Agriculture (Tractor Cabs) Regulation 1974 (Statutory Instrument 1974, No. 2034), almost all tractors must be fitted with an approved cab or frame which will prevent injury in the case of the tractor overturning. Since 1977 all cabs must be designed to keep the level of noise inside to below 90 dB(A) (Health and Safety Executive, 1978).

Power take-off (pto) shafts The Agricultural Power Take-off Regulations, 1957 (Statutory Instrument 1957, No. 1386) state that while the engine is in motion the pto must be covered by a shield which will prevent contact from above or from either side and will support at least 250 lb (113.4 kg), or that the pto must be enclosed by a fixed cover. Shafts must be enclosed, while in motion, in a guard which extends the whole length from the tractor to the first bearing on the machine.

Tractor Mounted Equipment

When coupling a tractor it should be backed up to the implement, stopped, put into neutral and the brake applied. The driver should then dismount and couple up with both his feet on the ground. When using trailers or trailed implements the towing pin should be adequate and secured with a nut or cotter pin. People should never be carried on a trailed implement unless it is fitted with a proper seat or standing platform. The use of any implements in this category is covered by the Agriculture (Field Machinery) Regulations 1962 (Statutory Instrument 1962, No. 1472).

Machine guards Machinery can maim or kill and all moving parts must be adequately guarded, also any driving belts or chains that have fastenings that might injure a worker must be guarded along their whole length. A device, readily accessible to the operator, must be fitted that will quickly stop any prime mover (the moving part of the power source), and that when set in the 'off' or 'stop' position will prevent the prime mover being started.

Drawbar Unless the drawbar is designed exclusively for attachment other than by manual means, every drawbar to a two-wheeled trailer whose unladen weight is more than 10 cwt (508 kg), or to any other two-wheeled machine when the downward force exerted by the drawbar at the point of the hitch is more than 56 lb (25 kg), must be fitted with a jack for raising or lowering the drawbar (The Agriculture (Field Machinery) Regulations) (Statutory Instrument 1962, No. 1472).

Brakes Trailers must be fitted with a separate braking system; this must be at least a hand-operated device which can be reached by the driver of the tractor.

Self-propelled Equipment

Every self-propelled (propelled by its own power source) and pedestrian-operated machine must be fitted with a stop switch which is easily accessible to the driver in his normal position. In addition to this, every self-propelled implement, except those controlled by pedestrians, must be fitted with a device by which the power from the prime mover can be disconnected quickly. Both these devices must be clearly marked (Health and Safety Executive, 1978). When operating pedestrian-controlled implements such as cultivators or cutters it is advisable to wear steel-capped safety footwear.

Spraying Equipment

All spraying equipment should be adequately maintained and tested.

PESTICIDES AND HERBICIDES

Poisonous Substances

It should be remembered that effective pesticides are necessarily toxic to some form of life and care should be taken when handling them. Wherever possible, a product approved by the Ministry of Agriculture, Fisheries and Food (1982) should be used. Some substances are controlled by the Health and Safety Poisonous Substances in Agriculture Regulations 1984 (Statutory schedules to this Instrument which specifies how they can be used, what precautions have to be taken, and excludes persons under the age of 18 years from carrying out scheduled operations with specified substances). Some pesticides are also covered by the Poisons Rules 1978 (Statutory Instrument 1978, No. 1), which lists the restrictions on sale, supply, labelling, precautions, containers, storage, and transport.

Labels should always be read and instructions followed implicitly. If chemicals have to be transferred to another container it must always be adequately labelled. A bottle which has previously contained beverages such as milk or soft drinks must never be used in case it is thought still to contain them. It is illegal to transfer scheduled poisons into containers which do not carry an approved label. Care should be taken to avoid getting pesticide on to the skin, into the mouth or the eyes, and to avoid inhaling any vapours, dusts or spray droplets. It should be remembered that pesticides are most dangerous in their concentrated form. But even after a crop has been sprayed it may present a hazard and great care should be taken to avoid contact. No part of the crop should be eaten until the recommended time has elapsed since spraying. This also applies to wild fruits in hedgerows.

There are several publications about the use of pesticides and other poisonous substances on farms (American Public Health Association, 1967; National Institute for Occupational Safety and Health, 1976; Health and Safety Executive, 1980; British Crop Protection Council, 1983). These should be available for reference.

Protective Clothing

Always wear the protective clothing described on the label and where scheduled compounds are being used make sure that the protective clothing you have chosen complies with that specified in the regulation. Protective clothing can exacerbate heat stress (refer to section on heat exhaustion under glasshouses and cloches), and it must be decontaminated after use (see below).

Precautions and First Aid

Washing It is important that any accidental splashes or spillages of chemical can be washed from the skin; it is therefore advisable (and with scheduled substances obligatory) to carry into the field the following items: soap, a towel and a supply of clean water in containers clearly marked 'personal washing'. It is also useful to carry a change of clothes.

First aid A first-aid kit should always be easily available. It should not only contain the basic essentials for first aid (The Health and Safety (First Aid) Regulations 1981), but also a disposable container which holds at least 300 ml of sterile eye wash solution, and the antidotes (if available) for the chemicals that may be handled. If any contamination occurs, a doctor should be consulted after rendering first aid.

Disposal of waste material Under the Control of Pollution Act 1974 it is an offence to deposit on land any waste which is likely to cause pollution, and special care should be taken to avoid pesticide entering streams or ditches. It is advisable to take any empty containers or unwanted pesticide away for disposal in a manner approved under the Act.

Decontamination After finishing a field trial it is important that all protective clothing that has been used is decontaminated before it is stored for subsequent re-use. The product label will normally give details of how to decontaminate. Whenever possible preliminary decontamination should be carried out in the field with more thorough decontamination back at base. Thorough decontamination of person and clothing must always precede eating, drinking, and smoking.

GLASSHOUSES AND CLOCHES

Glasshouses and cloches are constructed of glass which is easily broken to produce very sharp edges which can cause deep cuts. Care should therefore be taken when working inside glasshouses or in areas where cloches are erected, especially if extended spraying lances or booms are being used. Some glasshouses contain mercury-vapour lights which can implode, and eye protection should be employed.

Spraying and Fumigation

Spraying inside or fumigating a glasshouse creates an abnormal chemical atmosphere inside an enclosed space. Protective clothing including suitable respiratory protection must always be worn. Spraying should be done while walking backwards towards the door, with care taken not to lose the way. On completion the door should be locked, with a notice on it indicating that spraying or fumigation has taken place. This notice should say which chemical has been sprayed and specify when it is safe to enter without taking precautions.

Heat Exhaustion

In warm weather the temperature inside a glasshouse can rise rapidly to very high levels, and when the period of work there is prolonged, especially if protective clothes are being worn, it is advisable that plenty of fluid be drunk and salt tablets taken to guard against heat exhaustion.

REFERENCES

Agriculture (Field Machinery) Regulations. (Statutory Instrument 1962, No. 1472).
Agriculture (Power Take-off) Regulations. (Statutory Instrument 1957, No. 1386).
Agriculture (Tractor Cabs) Regulations. (Statutory Instrument 1974, No. 2034).
American Public Health Association (1967). *Safe Use of Pesticides.* New York: American Public Health Association.
British Crop Protection Council (1983). *The Pesticide Manual,* 7th edn. Croydon: BCPC.
Control of Pollution Act (1974). London: HMSO.
Health and Safety Executive (1978). *A Guide to Agricultural Legislation.* London: HMSO.
Health and Safety Executive (1980). *Poisonous Chemicals on the Farm.* London: HMSO.
Health and Safety (First Aid) Regulations. (Statutory Instrument 1981, No. 917).
Institute of Biology (1980). *Safety in Biological Fieldwork—Guidance Notes for Codes of Practice.* D. Nichols (ed.). London: Institute of Biology.
Ministry of Agriculture, Fisheries and Food (1982). *Approved Products for Farmers and Growers.* London: HMSO.
National Institute for Occupational Safety and Health (1976). *Safe Use of Pesticides.* USA Department of Health, Education and Welfare Publication No. (NIOSH) 76-147.*
Poisons Rules. (Statutory Instrument 1978, No. 1).
Poisonous Substances in Agriculture Regulations. (Statutory Instrument 1984, No. 1114).

---

* Available from Microinfo Ltd, PO Box 3, Hamlet House, Alton, Hants GU34 1EF.

8 Planning biological laboratories
K. Everett

Well-planned laboratories are less hazardous places in which to work than are those that are ill-conceived. Work in biological laboratories may be placed in three broad groups: teaching; research and development; analytical services. Each of these requires a different approach in planning individual rooms and groups of rooms. In teaching laboratories the students usually sit facing the lecturer and the benches are arranged in rows accordingly. For the other types of work a different arrangement, with island or peninsular benches may be more convenient.

The design of school laboratories is dealt with in detail by Archenold *et al.* (1978). Further information on safety in school laboratories may be found in Everett and Jenkins (1980), and others (Department of Education and Science, 1977, 1978; Schools Council, 1974; Anon., 1974). General problems of laboratory design are dealt with by Everett and Hughes (1981), Collins (1983) and WHO (1983). As these publications are readily available only an outline of laboratory planning need be given here, to stress the principles involved. The desirability of cooperation between the users and the designers cannot be stressed too much.

CHOICE OF LABORATORY FURNITURE

Three principal factors need to be considered:

1. Appropriateness of the physical dimensions.

2. Suitability of materials of construction.

3. Ease of access for cleaning.

Where much of the work is done by seated operators the benches should be lower than where they work while standing; knee-space will be required which limits the number of under-bench cupboards.

The first stages of the design or reorganization of any laboratory are generally assisted by the use of a combination of questionnaires, flow diagrams (Figure 1) and zone charts (Table 8.1). The aim of the flow diagrams and questionnaire is to identify *all* materials and *all* persons likely to enter the area. The zone chart will enable various risk areas to be identified. The preparation of these items will be assisted by reference to the Statement of Safety Policy (required under the Health and Safety at Work etc. Act 1974) and any existing house rules and codes of practice. It is important to identify all peripheral activities as well as the central laboratory operations, for example, out-of-hours' work, maintenance, cleaning, portering, office activity, and visitors. The flow diagram (Figure 1), used as an example, is for an analytical laboratory which receives biological samples from remotely located customer departments. Each type of laboratory will have its own flow diagram which varies in detail from others.

HAZARD AND RISK ASSESSMENT

In the earliest stages of a design project or in a review of the management of an existing laboratory it is essential that current legislation and relevant codes of practice are studied. The (Howie) Code of Practice (Department of Health and Social Security, 1978) is a necessary guide to the categorization of a wide range of organisms, but note should be taken of the full title of the Code, which places some limitations on its application. The first report of the Advisory Committee on Dangerous Pathogens (HMSO, 1984) which supersedes parts of the Howie Code is essential reading. A further report may be expected to complete the process of replacement of the Howie Code.

A serious attempt must be made to assess both the hazards and the risks, as defined below, of the work. In this context: *hazard* is defined as the magnitude of the danger, and *risk* as the probability of the danger manifesting itself.

The possible consequences of an accident and its probability must be assessed realistically and the design of the building modified accordingly. It should also be borne in mind that the staffing arrangements and managerial controls in force may also influence design requirements.

The scale of operations should be identified and some attempt made to predict changes which will affect laboratory layout, for example, increasing workload, automation, increasingly sophisticated equipment.

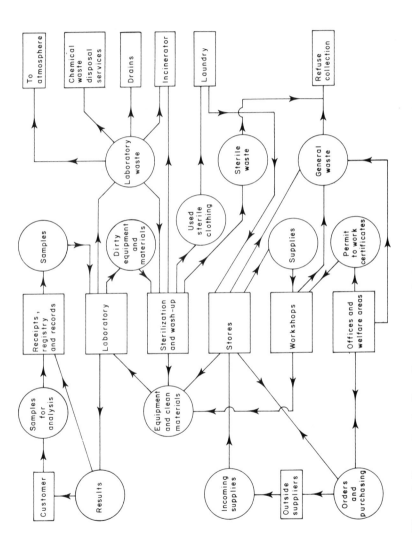

Figure 1. Example of a flow diagram: an analytic laboratory which receives biological samples

Table 8.1 Zone Chart to Identify Risk Areas

Red Zone (high-hazard areas)	Samples for analysis, receipts and registry area	High-hazard laboratory area	Dirty side of sterilization area	Hazardous materials areas
Orange Zone (medium-hazard areas)		Ordinary laboratories	Plant rooms	Wash-up areas
Green Zone (low-hazard areas)	Offices	Workshops Stores	Welfare areas	Library Seminar rooms	Clean side of sterilization area Public circulation areas

Allowance must be made for the fact that as limits of analytical detection are lowered so the sophistication of equipment and standards of hygiene and cleanliness tend to be raised.

The scale and nature of the operations in relation to the available site normally determine the actual layout, and idealized layouts are rarely possible in practice. Some effort should be made to group operations of similar hazard together so that there can be a progression from low- through medium- to high-hazard areas. If this can be achieved ventilation systems, for example, can be designed which draw air into the low-hazard zone and discharge it from the high-hazard zone.

ANIMAL ACCOMMODATION

The problems of safety in the animal house are well reviewed by Seamer and Wood (1981) and much of the advice given may be usefully translated into the context of laboratories using biologically active materials. Considerable attention must be paid to security and Home Office licence requirements. See also Chapter 2 and the Advisory Committee on Dangerous Pathogens (1984).

GENERAL ACCESS

The provision of ready access for the delivery of goods and samples, and for emergency services is important, as is adequate parking for staff and visitors. The question of the future use of motor-cycles and bicycles should also be considered, particularly when there is a high proportion of junior staff.

FIRE PRECAUTIONS AND SOLVENT STORAGE

At an early stage of the design the advice of the local fire authority should be obtained and the layout of the complex should take account of both fire prevention and fire fighting needs (Everett and Hughes, 1981). Adequate and suitable storage units for flammable solvents must be provided. When calculating the amount of solvent storage capacity note must be taken of the need to provide space for waste solvents being held prior to disposal.

GAS CYLINDERS

At the time of writing the legislation about the storage and use of gas cylinders is under review. The design of buildings, particularly multi-storey buildings, must take account of the need to move gas cylinders and other large or heavy equipment from time to time. The high proportion of slightly-built female staff employed in laboratories must also be considered.

It would appear that pressure will be exerted for the placing of cylinders out-of-doors and for the gas to be piped to the point of use. In particular, liquefied flammable gases (propane, butane, LPG, etc.) and acetylene are the subject of review and new statutory regulations are in preparation. It is strongly advised that the local Health and Safety Inspectors are consulted on this point.

GENERAL HYGIENE

In biological laboratories a generous provision of hand-washing facilities should be made, any available codes of practice being used as guidance (Everett and Hughes, 1979; Everett and Jenkins, 1980; Advisory Committee on Dangerous Pathogens, 1984).

It is essential that adequate provision should be made for clothing lockers, both outdoor and laboratory wear, tea and rest rooms, smoking areas, etc.

First-aid arrangements should be simple and well understood. Where biologically active materials are in use *all* injuries and accidents must be reported and recorded promptly and an appropriate system must be available to encourage staff to cooperate for their own benefit. Attention is drawn to the statutory requirement to notify accidents and dangerous occurrences (HSE, 1980).

LABORATORY LAYOUT

When designing a laboratory layout it is important to distinguish between general trends and needs and personal idiosyncrasies. The design must be as flexible as possible to take account of staff changes and also of developments in the science. It is important to try and identify items of equipment which have a potentially long life (say, 20 years or more) and those with a fairly short life expectancy (say, 10 years or less). An example of the former group might be a fume cupboard or an electron microscope, the latter group might include a much-used centrifuge or an automatic analyser.

The design of easily modified laboratories has been the subject of considerable study and the concept of providing services independently of benches has much to commend it.

REFERENCES

Advisory Committee on Dangerous Pathogens (1984). *Categorisation of Pathogens According to Hazard and Categories of Containment.* London: HMSO.

Anon. (1974). *Precautions against Biological Hazards.* London: Imperial College of Science and Technology.

Archenold, W. F., et al. (1978). School science laboratories. *A Handbook of Design, Management and Organisation.* London: John Murray.

Collins, C. H. (1983). *Laboratory-acquired Infections.* London: Butterworths.

Department of Education and Science (1977). *The Use of Micro-organisms in Schools.* Education pamphlet No. 61. London: HMSO.

Department of Education and Science (1978). *Safety in Science Laboratories.* DES Safety Series No. 2, 3rd edn. London: HMSO.

Department of Health and Social Security (1978). *Code of Practice for the Prevention of Infection in Clinical Laboratories and Post-mortem Rooms.* London: HMSO.

Everett, K., and Hughes, D. (1979). *A Guide to Laboratory Design.* Reprinted with further additions. London: Butterworths.

Everett, K., and Jenkins, E. W. (1980). *A Safety Handbook for Science Teachers,* 3rd edn. London: John Murray.

HSE (1980). The Notification of Accidents and Dangerous Occurrences Regulations 1980. Health and Safety. No. 804. London: HMSO (under review).

Schools Council (1974). *Recommended Practice for Schools Related to the Use of Living Organisms and Material of Living Origin.* London: English Universities Press.

Seamer, J. H., and Wood, M. (eds) (1981). *Safety in the Animal House. Laboratory Animals Handbooks 5,* 2nd edn. London: Laboratory Animals.

World Health Organization (1983). *Laboratory Biosafety Manual.* London: HMSO.

9 Safety in the use of chemicals

F. Grover

To some extent the increasing specialization which has taken place in recent years has increased the risks of accidents arising from the uninformed use of chemicals and other substances hazardous to health. The principal objective of this book is to alert biologists and others to the hazards which they may encounter in the accomplishment of their daily tasks.

The key to the safe use of chemicals in the laboratory is the availability of accurate information about their essential properties. Before any chemical is used, it is important that the worker is fully aware of any risks associated with the substance and that he has all the equipment necessary to minimize any potential risks. The ultimate disposal of hazardous chemicals must also be considered. Too many fume cupboards in chemical laboratories are cluttered with old stocks of hazardous chemicals, and sometimes with the products of failed reactions that no one in the laboratory knows how to get rid of.

CONTROL OF SUBSTANCES HAZARDOUS TO HEALTH (COSHH REGULATIONS)

Since 1974 the Health and Safety at Work Act has placed a general duty on employers to safeguard, as far as is reasonably practicable, the health, safety, and welfare at work of all their staff. There is also an obligation to protect the public from the employer's acts or omissions, and in addition the employee has a duty to take reasonable steps to protect himself. If the proposed 'COSHH' Regulations, which have recently been published for comment, are accepted, the employer or his line management team will have to assess all substances hazardous to health before they are put into use (Health and Safety Commission, 1984). If this assessment is to be carried out properly it will be imperative that adequate information is available to staff and that due account is taken of the hazards to be encountered in the use of laboratory chemicals.

INFORMATION REQUIRED

Before any chemical is used we need to know something of its properties. In what is it soluble or insoluble? Is it flammable, explosive or otherwise highly reactive? Are there special problems concerning its incompatibility with certain groups of chemicals? Is it toxic, carcinogenic, mutagenic or teratogenic? Is it lachrymatory or otherwise harmful to human tissue? If it has any of these properties, it is important to have information concerning its volatility and vapour pressure at room temperature. Finally, how will it be disposed of? Can it go down the laboratory sink? Will a particular treatment make it innocuous or less harmful? Or will a contractor have to be paid to take it away?

SOURCES OF INFORMATION

Apart from the standard chemical textbooks, some manufacturers' and suppliers' catalogues are valuable sources of information. The larger general suppliers nearly always give details of the chemical formula, molecular weight, boiling point, and in many cases will indicate if a substance is poisonous or comes within the purview of the Dangerous Drugs Regulations. One catalogue also suggests methods of disposal for each of the 12 000 substances listed (*Catalogue/Handbook of Fine Chemicals*, 1985).

In addition there are standard reference books which are generally available in science libraries and which give details of the properties of a very wide range of chemicals (Merk Index, 1983; National Institute of Occupational Safety and Health, 1983; Irving, 1984; Bretherick, 1985).

CONSIDERATION OF THE HAZARDOUS PROPERTIES OF CHEMICALS

Solubility

If a substance is soluble in water, then there is an obvious and readily available source of dilution which may make its safe use and ultimate disposal that much easier. It should be borne in mind, however, that some chemicals may react extremely violently with water and cannot be treated in this way.

Clearly, extremely toxic substances cannot be put down the laboratory drain, but if a water-soluble compound catches fire, the fire may be put out, or the risk of fire from an exothermic reaction may be reduced, by the addition of water. In general, solvents immiscible with water cannot be put into the drains. Some substances, e.g. sodium hydroxide, dissolve in water with the evolution of a large amount of heat and may cause a reaction to run out of control. With others the converse is true and heat will be extracted from the reaction.

Flammability and Explosibility

If a solid substance will burn, it is necessary to have some idea of its rate of combustion in air, and to know what is likely to initiate combustion. It is also useful to have some idea of the products of combustion.

In the case of a flammable liquid, reference to its flash point and auto-ignition temperature will provide a useful indication of the precautions which need to be taken to prevent a fire. For example, some liquids such as carbon disulphide, have an extremely low flash point and, it is said, may be ignited by a hot steam pipe, while others require a much higher temperature to initiate combustion.

Gases and vapours may also have upper and lower explosive limits in air and this should be kept in mind when assessing the safety of an experiment. Obviously a substance which will explode when present in air at relatively low concentration will require additional precautions to prevent its escape either as a gas, or vapour. Substances which are capable of exploding over a wide range of concentrations, e.g. diethyl ether (1.85 - 48 per cent) and hydrogen (4 - 75 per cent) will clearly need to be treated with rather more respect than, say, toluene which has explosive limits of 1.4 - 6.7 per cent or ammonia which has explosive limits of 16 - 25 per cent.

The vapour density of flammable solvents should also be considered. Diethyl ether has a very heavy vapour which if released in the laboratory will fall into the gaps between benches and may produce a dangerous concentration sufficient to cause an explosion if the means of ignition are present. If the escape of flammable gases or vapours cannot be prevented, then the experiment should be carried out in an efficient fume cupboard.

Some chemicals will explode or detonate (the difference between the two being largely the rate of reaction) if struck in the dry state. Sodium azide is widely used in biological preparations as a preservative and even dilute solutions, if they are allowed to come into contact with metals, will produce metal azides which are capable of exploding with extreme violence. Explosions have been recorded where biological solutions preserved with sodium azide have been freeze dried under vacuum and where hydrazoic acid has been released which has passed through various traps and subsequently reacted with copper pipework to form copper azide which has subsequently exploded (DHSS, 1982).

Incompatibility

Some substances which appear to be relatively inert may, under appropriate conditions, react violently or in an unexpected manner if they are brought together either inadvertently or deliberately in the course of a preparation. Examples of this are the reactions of magnesium with chloroform, and chloroform and acetone. In some circumstances formaldehyde solution will react with hydrochloric acid or sodium hypochlorite solution to produce bischloromethyl ether which is an extremely toxic and carcinogenic agent.

A useful reference book here is that of Bretherick (1979) who has made a particular study of reactive chemical hazards. This book will be found most useful to safety officers and others who have a particular responsibility for chemical safety.

TOXICITY

Many chemicals which we encounter in our daily lives either at work or at home, are toxic if fed to man in sufficient quantity. The question is how toxic? For many years the tables compiled by the American Conference of Governmental Industrial Hygienists (ACGIH) have been used to assess the health risks to those working with a wide range of chemical substances. Threshold Limit Values (TLVs), expressed as parts per million (ppm) of the substance in air, gave a considered view on the maximum allowable time-weighted average exposure for staff working an 8-hour day, 5-day week. Some limits were expressed as mg/m^3 or in the case of some fibrous materials as fibres per ml. In addition to their TLVs, some substances were given 'ceiling limits' (CLs) which were maximum levels, or 'short-term limit values' (STLVs) which were short-term averages with only limited excursions allowable above the given norm.

In the United Kingdom, ACGIH-derived Threshold Limit Values have been superseded by 'Occupational Exposure Limits' published as a Guidance Note by the Health and Safety Executive (1984). We now have recommended limits for an 8-hour time-weighted average (long-term exposure) and for a 10-minute time-weighted average (short-term exposure). There is also a short table of 'control limits' for a certain range of substances where specific limits have been set by 'Regulation', 'Approved Codes of Practice', 'European Community Directives', or which have been specifically adopted by the Health and Safety Commission.

Long-term exposure limits range from 5000 ppm for substances such as carbon dioxide, through 1000 ppm for acetone, ethanol, and liquefied petroleum gas, down to 0.0002 ppm for osmium tetroxide. These limits are only in respect of the toxic properties of the substances mentioned. Liquefied petroleum gas, for example, is not particularly toxic, but its flammable and explosive properties are likely to be far more hazardous.

The values given in the Occupational Exposure Limits tables are intended to represent *maximum* limits of exposure. Wherever possible as low a level as is reasonably practicable should be achieved. For persons with limited experience, and provided that exposure limits are not considered in isolation, these tables are a readily available, inexpensive, and easily understood source of information on toxicity. Other sources of reference information have been quoted above.

CARCINOGENIC, TERATOGENIC, AND MUTAGENIC AGENTS

Definitions

Carcinogen   A substance agreed by competent authorities to cause cancer in man or animals at a specified level of exposure. The incubation period may be relatively short or as long as 30 to 40 years (MRC, 1981).

Suspect carcinogen   A chemical, not itself known to be a carcinogen but with a close structural similarity to a known chemical carcinogen and/or for which there is preliminary or unconfirmed positive evidence from animal tests or epidemiological studies (MRC, 1981).

Teratogen   An embryotoxin that induces foetal abnormalities at concentrations below those causing embryonic death (Hartree and Booth, 1977).

Mutagen   A substance that induces irreversible chemical aberrations in chromosomal nucleic acid (DNA) (Hartree and Booth, 1977).

The possibility that a particular substance might possess one or more of these properties should always be considered before it is used, particularly for the first time. In general, man does not immediately collapse or exhibit other untoward symptoms if exposed to these substances: the effects are usually much more insidious. The person exposed may not become aware of the potential hazard, nor appreciate the devastating effects of lax methods of work until many years later. It is important, therefore, that all staff should follow an agreed code of practice and adopt a high standard of personal hygiene whenever they work with these substances.

For work with substantial amounts of potent carcinogens the Medical Research Council's (1981) *Guidelines for Work with Chemical Carcinogens* gives useful advice, including a model code of practice which can be adapted to suit the particular circumstances in a wide range of laboratories.

At a different level, *Precautions for Laboratory Workers Who Handle Carcinogenic Aromatic Amines*, published by the Institute of Cancer Research (ICR, 1966), gives relatively simple and easy-to-apply recommendations which can be adapted for use in almost any type of laboratory where carcinogenic or toxic chemicals are handled.

Apart from the obvious advice on the labelling and storage of carcinogens, the use of protective clothing and the avoidance of skin contact is also discussed in these booklets. Regrettably the necessity of adopting these precautions has not always been sufficiently appreciated in some laboratories. The following points are emphasized in the ICR monograph and could well form the basis of a code of practice for any work with toxic chemicals.

1. If there is any accidental contact, the affected parts (including the eyes) should immediately be washed in *cold* running water for 5 minutes.

2. Any operations involving the risk of vapour or dust formation should be carried out in a properly exhausted fume cupboard.

3. Overalls should always be worn and should be changed after the experiment is completed. Rubber or plastic gloves, when used, should be well washed in cold running water.
4. Empty apparatus and storage jars should first be rinsed out in cold water after use. Benches should be washed with *cold* water.
5. Even when these precautions have been followed, hands should be well rinsed after using these substances with *cold* water first, before soap is used.
6. Impervious working bench surfaces are essential (there may be absorption into wood surfaces).

All these precautions are sensible and easy to apply. They could be profitably followed whenever work with hazardous chemicals is contemplated. In summary, the simple advice is: (a) avoid all exposure either by inhalation, ingestion or absorption; (b) use the appropriate protective clothing and equipment; and (c) if the skin is contaminated remove the contamination with cold water as soon as possible.

For the disposal of carcinogens, teratogens, and mutagens, chemical destruction (provided that the appropriate facilities and expertise are available) is the preferred method. Incineration may also be considered but is not advised for volatile materials. It is important to remember that the gaseous products of combustion, or alternatively the residues of incomplete combustion, may also be hazardous and steps must be taken to prevent their escape into the environment.

USE OF CHEMICALS: GENERAL CONSIDERATIONS

All chemicals must be handled with regard to their chemical and physical properties. It must be emphasized that in all laboratories where toxic substances are used there should be no mouth pipetting, eating or drinking, smoking, application of cosmetics, or sucking of pencils or licking of labels.

In dealing with chemical hazards the following general precautions should be considered:

1. Full information should be available before the experiment is started.
2. The smallest practicable amounts of substance should be used.

Table 9.1 Comparative Properties of some Commonly Used Solvents

Solvent	Long-term Exposure Limit		Control Limit		Boiling Point (°C)
	(ppm)	(mg m^{-3})	(ppm)	(mg m^{-3})	
Acetone	1000	2400	—	—	56
Benzene	10	30	—	—	80
Carbon disulphide	—	—	10	30	46
Carbon tetrachloride	10	65	—	—	77
Chloroform	10	50	—	—	61
Diethyl ether	400	1200	—	—	34
Ethanol	1000	1900	—	—	79
Hexane	100	360	—	—	69 approx
Methanol	200	260	—	—	65
Toluene	100	375			111
Xylene	100	435	—	—	140 approx

Flash Point (°C)	Ignition Point (°C)	Explosive Limits in Air (%)	Water Miscible	Remarks
−18	538	3–13	Yes	Highly flammable
−11	562	1.4–8	No	Highly flammable. Toxic by inhalation and skin contact
−30	100	1–44	No	Extremely flammable. Very toxic by inhalation Causes severe damage to the nervous system
−	−	−	No	Very toxic by inhalation and skin contact
−	−	−	No	Harmful by inhalation. Will defat the skin
−45	180	1.85–48	No	Extremely flammable. May form explosive peroxides
12	423	3.3–19	Yes	Highly flammable
−23	260	1.2–7.5	No	Highly flammable
10	464	7.3–36.5	Yes	Highly flammable. Toxic by inhalation or if swallowed
4.4	536	1.4–6.7	No	Highly flammable. Harmful by inhalation Absorbed through skin
17–25	464–529	1–7	No	Flammable. Harmful by inhalation

3. The less toxic rather than more toxic substance should be used.
4. Non-flammable or less flammable solvents should be used whenever possible.
5. Appropriate protective clothing (gloves, overalls, goggles, boots, etc.) must always be available and be used when necessary. Gloves in particular, must be free from pinholes and made of material known to be impervious to the substances being used. After use this clothing, etc., must be washed, decontaminated or discarded before the worker enters uncontaminated areas and certainly before visiting restaurants or offices. Laboratory overalls are provided for protection: they must not be seen as a 'badge of office' in public areas.
6. In general, unless large volumes of aerosols are likely to be generated, toxic powders are more hazardous than toxic solutions. Just as microbiologists use microbiological safety cabinets as their first line of defence so should chemists or others who use hazardous chemicals consider the fume cupboard as their prime means of protection. The use of personal respiratory equipment should normally only be a measure of last resort.

Where there is a choice between particular substances, one of which is flammable and another which is not, then the choice is clear. The difficulty is with distinguishing between one substance which is flammable but non-toxic and another which is toxic and non-flammable. The only real solution to this type of dilemma is to have all the available information and to attempt to relate the precautions required to the facilities and equipment available in the laboratory concerned (see Table 9.1).

STORAGE OF CHEMICALS

The essence of good chemical storage is: (a) minimum practicable amounts; (b) segregation; and (c) suitable conditions (Everett and Hughes, 1979; Grover and Wallace, 1979). In general, bulk flammable solvents must be stored in specially constructed solvent stores. In the case of petroleum products, licensed 'petroleum stores' may have to be provided. It is also important that strong acids are stored away from strong alkalis, and substances which are strongly oxidizing away from those which are powerful reducing agents. The temperature and ventilation of the store-room are equally important and no highly reactive, corrosive or other material which is particularly hazardous should be kept on high shelves.

Ordering and maintenance of chemical stocks is a task which requires some technical knowledge. In one example which has been well cited in the literature, the store-keeper who had very little technical knowledge and who was required as soon as possible to organize a new general chemical store, ordered a dozen Winchester quarts of everything he could recognize in the suppliers' catalogue, including 12 bottles of 'duty paid' ethyl alcohol and 12 bottles of 70 per cent perchloric acid!

Finally, in any well-organized chemical store, the stock should be issued in the same order in which it was received from the supplier. A recent Hazard Note (DHSS, 1985) mentioned that following a series of explosions, hospital laboratory stores were inspected and one bottle, containing a mixture of diethyl ether and hydrogen peroxide (ozonic ether), was estimated to be between 27 and 32 years old! As peroxides had been found in other stocks of diethyl ether in the same laboratory, this particular bottle was considered to be so hazardous that staff from the National Chemical Emergency Centre at Harwell had to be commissioned to ensure its safe removal and disposal.

DISPOSAL OF WASTE CHEMICALS

The best method of avoiding problems with disposal is, if at all possible, not to use substances which will be difficult to dispose of. It is also important to avoid the accumulation of large quantities of hazardous waste which may make its eventual disposal more difficult. The removal of small amounts of any hazardous chemical by outside contractors is expensive and in some cases prohibitively so. It is better, if at all practicable, to render such material harmless by chemical transformation or incineration on site.

If hazardous waste has to be stored pending its ultimate disposal then the normal rules of storage for chemicals will apply. It must be put into clean containers, segregated, and stored under the appropriate conditions. On no account should waste chemicals, particularly solvents, be mixed unless it is known that the resultant mixture will be safe.

In general, no immiscible solvent may be put into the drains and it should be noted that all local authorities have regulations and by-laws which control the types and quantities of substances that can be put into the drains or dumped at controlled tips. These regulations are usually rigorously enforced and this, together with the application of the Control of Pollution (Special Waste) Regulations 1980, will make it increasingly difficult and expensive to dispose of hazardous waste.

Specialized waste disposal contractors may be an expensive but relatively easy solution to the problem of hazardous or toxic waste disposal. Their special knowledge and experience in completing the necessary documentation certainly relieves the user of much of the worry, but there are pitfalls into which the unwary may fall. In recent times, some of these contractors have not exhibited the care and concern which a reasonable client would expect. It is the responsibility of the employer and of his staff to ensure that the contractor is exercising his responsibilities correctly.

Waste flammable solvents, provided that they do not give off toxic products of combustion, may sometimes be used for demonstrations of firefighting techniques, but this must only be done under strictly controlled conditions. Certainly an experienced fire officer must be in attendance and the site and weather conditions must be suitable.

FUME CUPBOARDS: THEIR USE AND DESIGN REQUIREMENTS

Fume cupboards range from simple boxes with holes in the top through which the exhaust is extracted and which need a considerable volume of air-flow to give adequate containment, to sophisticated aerodynamic designs which are capable of good and uniform containment without removing unnecessarily large volumes of air from the laboratory. Many fume cupboards are inadequate for use with very toxic substances, and it is essential that their containment performance is known.

For some time the British Standards Institute (BSI) has been considering the publication of a British Standard for laboratory fume cupboards. A 'Draft for Development' DD80 was released and comment sought from users, suppliers, and installers (BSI, 1982). Part 1 of the draft deals with safety requirements and performance testing; Part 2 deals with the information to be exchanged between the purchaser, vendor, and the installer; and Part 3 makes recommendations for the selection, use, and maintenance of fume cupboards. It is Part 3 which is likely to be of most interest to the average user.

Unless an unfortunate choice is made with material of construction, a fume cupboard is expected to last for some years. It is important therefore that its design, construction, installation, maintenance, and performance are fully appreciated before an order is placed.

With the development of microbiological safety cabinets there has been a tendency in some quarters to confuse safety cabinets with fume cupboards. Clearly there are similarities: both are containment facilities which depend on an air extraction system to protect the operator from the hazardous materials or substances with which he works, but there the similarity must end. The safety cabinet is fitted with HEPA (high efficiency particulate air) filters, which will reliably remove bacteria, viruses, and other particulate matter from the exhaust but will allow volatile and gaseous substances to pass straight through. On the other hand fume cupboards are rarely fitted with either absorption or particulate filters and for various reasons they must not be used for microbiological work.

Given that a fume cupboard installation may be in service for many years and may therefore have to be used for a wide range of different purposes it is important that the following points are considered.

1. Its construction must be adequate to withstand the effects of the full range of chemicals likely to be used in it.

2. Its containment (i.e. its ability to capture and retain volatile materials) should be sufficient to protect the operator from the most toxic substances likely to be used.

3. The amount of air extracted, together with the projected rate of release of toxic substances, must be sufficient to ensure adequate dilution and dispersal so that no person downwind of the exhaust outlet will be placed at risk from the toxic effects of the effluent.

Choice of Material

Much will depend on the type of work to be carried out. Some grades of stainless steel are useful where radioactive chemicals are used because they are readily cleaned and/or decontaminated. If large numbers of acid digestions are to be carried out, however, stainless steel will be found to have a very short life indeed. Some painted or powder-coated metals do have a reasonable life-span with corrosive vapours, but much will depend on the protective coatings' freedom from 'pinholes' and the care with which the cupboard is used and maintained.

Provided that there are no traps into which condensed vapours may drain, the choice of material used for the duct work may be of less importance. Welded PVC duct of circular cross-section is widely used, but surprisingly, good results are obtained with coated (acid-proof paints) or bituminized iron.

Fibre glass has recently been used as a lining material for fume cupboard carcasses and although considerable doubts were originally expressed concerning its fire resistance, some makers are now claiming 'Class 1' or even 'Class 0' fire resistance. Generally it has good chemical resistance and although some types will not stand immersion in solvents, they will certainly withstand minor spillages or exposure to solvent vapours.

A few users will need fume cupboards which are able to withstand particularly corrosive substances. Large volumes of strongly acid fumes may require a water scrubber in the extract system, but this needs careful design and evaluation otherwise the only result may be to generate a dilute acid spray which falls on to the nearby roof! Perchloric acid and perchloric/nitric acid digestion systems will impose a particularly difficult requirement on the fume cupboard lining and exhaust system. It is vital that nowhere, either in the fume cupboard, in the duct or in the fan should acid fumes be able to collect or be absorbed. Some modern installations for this type of work not only have water-wash scrubbers in the duct, but are provided with a continuous wash-down in the extract system. It is also important in these circumstances that no metal equipment is left in the fume cupboard and that the whole of the interior is regularly cleaned.

CONTAINMENT OF TOXIC SUBSTANCES

Because a fume cupboard is an open-fronted system it can never provide 'absolute' containment. Substances which are so toxic that nothing can be allowed to escape must be handled in closed systems. For slightly less hazardous substances various methods have been proposed for measuring the containment provided by a particular installation. These usually depend on the controlled release of a specified gas inside the cupboard and the measurement outside (usually in the vicinity of the operator's face) of the proportion which escapes. The major disadvantage with these techniques is that with an efficient installation, only a very small proportion of the gas released will escape and that the equipment used to detect the escaping gas may not be sufficiently sensitive to give a meaningful result. In these circumstances large volumes of fairly expensive gas may have to be released over a relatively long period to bring the measurement 'on scale'.

Recently it has been proposed that one of the systems used to measure the containment of microbiological safety cabinets could be easily adapted to measure fume cupboard containment. Although this may be a matter of controversy in some quarters, the KI-Discus method (Watkins and Watkins Ltd, Wareham, Dorset), which releases and measures the escape of a very fine aerosol of potassium iodide solution from the fume cupboard, has been widely used in MRC laboratories to compare the performance of various types of fume cupboard and to ensure that they have been properly installed.

FACE VELOCITY

Hughes (1980) has compared the recommended face velocity of various authorities and has found that they range from 0.25 m s^{-1} (50 ft min^{-1}) to 1.0 m s^{-1} (200 ft min^{-1}). Clearly this is also a matter of some controversy and pending the publication of more formal requirements, 0.5 m s^{-1} should be regarded as the minimum for most work with toxic chemicals. It is important that the air-flow through the aperture is reasonably uniform and that air being drawn into the front opening is not affected by adjacent walls, benches or equipment. It is also important that the operator makes no violent movements with his arms within the working space and that he, or others, no not unduly disturb the laboratory air patterns near to the fume cupboard in the course of their work.

'RECIRCULATING' FUME CUPBOARDS

Recirculating fume cupboards are containment facilities which at first sight may appear to be hybrids between microbiological safety cabinets and conventional fume cupboards. They are usually a little smaller than the average fume cupboard and may have a vertical or horizontal sliding sash. They are fitted with a pre-filter and one or more specialized absorption filters to suit the type of fume being released. In some circumstances an absolute (HEPA) filter may also be fitted to remove very fine particulate matter. There is a built-in fan which draws air through the front opening and then through the filters. The fumes are absorbed by the filters and the clean air is returned to the room.

There are clearly many potential advantages in the use of such a system.

1. It can be bought virtually 'off the shelf' and requires little or no installation. No duct work is required.
2. It is relatively portable and can easily be moved from one laboratory to another as the need arises.
3. In initial cost it is generally somewhat cheaper than a conventional fume cupboard.
4. As no air is removed from the laboratory, no costs are incurred in heating replacement air.

These are very considerable advantages but there are other considerations which should be taken into account.

1. Polymethyl methacrylate, which is generally used for the construction of the carcasses, is not a material of first choice where solvents or flammable substances are likely to be released.
2. The absorption filters used are said to be capable of retaining several litres of solvent. Is it really safe to have this quantity of hazardous material sitting around in the laboratory waiting for an accident to happen?
3. It is conceivable that two or more strongly incompatible substances might react together on the filter, or that one which has a greater affinity for the filter material might drive off a previously absorbed substance. There is little published evidence to counter these particular objections, and that which has been published is not very convincing.
4. Some filters are pre-treated with substances which are non-toxic and which have an objectionable smell that will 'bleed off' as the filter approaches capacity. Others have more sophisticated means of detecting 'bleed through' of the substances being released in the work area. One very recent design has a 'light-pipe' built into the filter which will detect a fire on the material and close a damper to prevent the fire spreading. All of these devices are useful safety features, but all except the first add considerably to the cost which may make the recirculating fume cupboard rather less competitive than would at first appear.
5. Although there may be considerable savings both in capital and in running costs, filters are not cheap and may have to be changed fairly often if the cupboard is much used.

Undoubtedly there is a place for recirculating fume cupboards in some laboratories, but their use needs to be considered with great care. Generally it may be better to use a conventional fume cupboard for the most hazardous work and to reserve this device for work where the fumes are a nuisance rather than potentially life threatening to staff. Certainly if they are used with toxic materials it must be known that the filters are capable of retaining *all* of the toxic effluent and there must be some reliable means of monitoring the filter's performance and its remaining ability to absorb.

REFERENCES

Bretherick, L. (1979). *Handbook of Reactive Chemical Hazards.* London: Butterworth.

Bretherick, L. (ed.) (1981). *Hazards in the Chemical Laboratory.* London: Royal Society of Chemistry.

Catalogue/Handbook of Fine Chemicals (1985). Gillingham, Dorset: Aldrich Chemical Co.

Chester Beatty Research Institute (1966). *Precautions for Laboratory Workers who Handle Carcinogenic Aromatic Amines.* London: Institute of Cancer Research.

Department of Health and Social Security (1982). *Danger of Explosions in Freeze-drying Equipment and Vacuum Systems Used to Process Materials Containing Sodium Azide.* HN (Hazard) 82.10. London: DHSS.

Department of Health and Social Security (1985). *Explosions in a Hospital Laboratory Fume Cupboard Due to Evaporation of Diethyl Ether Containing Peroxides.* HN (Hazard) 85-1. London: DHSS.

Everett, K., and Hughes, D. (1979). *A Guide to Laboratory Design.* London: Butterworth.

Grover, F., and Wallace, P. (1979). *Laboratory Organisation and Management.* London: Butterworth.

Hartree, E., and Booth, V. (eds) (1977). *Safety in Biological Laboratories.* London: Biochemical Society.

Health and Safety Commission (1984). *Control of Substances Hazardous to Health.* Draft Regulations and Draft Approved Code of Practice. London: HMSO.

Health and Safety Executive (1984). *Occupational Exposure Limits 1984.* Guidance Note EH40. London: HMSO.

Hughes, D. (1980). *A Literature Survey and Design Study of Fume Cupboards and Fume Dispersal Systems.* Leeds: Science Reviews Ltd.

Laboratory Fume Cupboards (DD80) (1982). London: British Standards Institution.

Medical Research Council (1981). *Guidelines for Work with Chemical Carcinogens.* London: MRC.

Merck Index (1983). 10th edn. Rathway, NJ, USA: Merk.

National Institute of Occupational Safety and Health (1983). *Registry of the Toxic Effect of Chemical Substances.* Washington, DC: US Government Printing Office.

Sax Irving, N. (1984). *Dangerous Properties of Industrial Materials.* New York: Van Nostrand Reinhold.

Index

Accident reporting, 21, 72
Acts of Parliament, *see under name of Act*
Aerosols, infectious, 43
Agriculture (Field Machinery) Regulations, 62
Allergies,
  animal related, 21, 36
  plant related, 37, 41
Amphibians, 35
Anemones,
  garden, 37
  marine, 34
Animal
  Pathogens, Importation Order, 5
  products, 23
  temperament, 13, 17 *et seq.*
Animals,
  Act, 8
  barriers around, 15
  caging, 16, 27
  choice of species, 17, 26
  diseases from, 18 *et seq.*, 60
  during school holidays, 28
  gentling, 18
  hazardous to man, 31 *et seq.*
  health, 18 *et seq.*
  housing design, 15 *et seq.*
  importation, 7, 19
  in classroom and laboratory, 11 *et seq.*
  infections of, 19
  injuries caused by, 9, 21
  killing and disposal, 23, 28
  law relating to, 7
  manipulation, 22
  pests and contamination, excluding, 14
  supply of, 25
  venomous, 33
  wounds from, treatment, 21
Annelida, 32
Arthropods, 34
  parasitic, 36

Bees,
  Act and Order, 8
  stings, 34
Biological safety cabinets, 51
Blister plant, 37
Breach of Statutory Duty, 8
Brucellosis, 61
Buildings, safety in, 8

Carcinogenic Substances Regulations, 6
Carcinogens, 79 *et seq.*
Caterpillars, 32
Celandine, 37
Cephalopods, 32
Chemicals,
  disposal, 85
  explosive, 76, 78
  flammable, 76
  general precautions, 81 *et seq.*
  incompatible, 78
  ordering, 85
  solubility, 76
  storage, 84
  toxicity, 78 *et seq.*
Children, hazards to, 19, 22, 61
Clean Air Act, 7
Cloches, 65
Cnidaria, 34
Codes of Practice, 24, 50, 68, 80
Collection of wild material, 27
Conservation of Wild Animals and Plants Act, 35

93

Control of Pollution,
  (Injurious Substances) Regulations, 6
  (Special Waste) Regulations, 6
Control of Substances Hazardous to Health Regulations, 75
Crustacea, 32

Dangerous
  chemicals, 75 *et seq.*
  Packaging and Labelling Regulations, 6
  Pathogens Regulations, 5
  Substances, 5
  Wild Animal Act 1977, 10
Decontamination, after spraying, etc., 65
Design
  animal houses, 15
  biological, 67 *et seq.*
  furniture, 67
  general hygiene, 72
  laboratories, 48 *et seq.* 67 *et seq.*
  layout, 72
  microbiological,
    basic, 48
    containment, 50
    maximum containment, 51
    school science, 48
Disinfectants,
  testing, 55
  use of, 15, 23, 53
Disposal of waste, 23, 52
  Acts and Regulations, 6
Driving, 60
Droplet nuclei, 43
Drugs, Misuse of, Act, 6
Dumping at Sea Act, 6

Electrical Equipment (Safety) Regulations, 27
Employers' liability, 8
Euphorbias, 38
Explosives Act, 6
Eye, infection through, 44

Farmer's lung, 41, 61
Field experiments, 59 *et seq.*
Fire
  precautions, 71
  Precautions Act, 5
  prevention, 5
First Aid arrangements, 72
Fish,
  hazardous, 35
  wounds from, treatment, 35
Fume cupboards, 86 *et seq.*
Fumigation, of glasshouses, 65

Gases, explosive and flammable, 77
Genetic Manipulation
  Advisory Group, 5
  Regulations, 5
Glasshouses, hazards in, 65, 66

Hazard Groups, of microorganisms, 46
Hazardous chemicals, 6
Hazchem, 5
Health and Safety Inspectors, 4, 24
Health and Safety at Work etc. Act, 1 *et seq.*
  duties imposed by, 2
  employers' policy documents, 3
  enforcement and liability, 4
  implementation of, 3
  legislation affecting biological laboratories, 4
Heat exhaustion, 66
Hygiene, 20, 50, 54

Ionising Radiations, Code of Practice, 5
Importation
  of plants, 7
  of wild animals, 7
Improvement Notices, 4
Infected material, disposal of, 23, 52
Infection,
  barriers against,
    primary, 45, 47 *et seq.*
    secondary, 45, 53 *et seq.*

factors involved, 44
general precautions, 54
routes of, 43
Ingestion, of microorganisms, 44
Inhalation, of microorganisms, 43
Injection, accidental, 44
Injurious Substances Regulations, 6
Invertebrates capable of inflicting injury, 31, 34 *et seq.*
Irritant plants, 37 *et seq.*

Jellyfish, 34

Kinetic handling, 60

Laboratory
  design, biological, 67 *et seq.*
  design, microbiological,
    basic, 48
    containment, 50
    maximum containment, 51
    school science, 48
  furniture, 67
  general hygiene, 72
  layout, 72
Laboratory Animal Breeders Association, 19, 25
Leeches, 32
Legal liabilities, in biology, 6 *et seq.*
  breaches of, 8
  insurance, 8
Lifting, 60
Living organisms, in classrooms, 28

Machinery,
  agricultural and horticultural, 61 *et seq.*
  guards, 62
  self-propelled, 63
Medical supervision, 54
Medicines Act, 6
  and poisons, 5
  (Veterinary Drugs) Orders, 6
Microbiological
  laboratory equipment, 52
  safety cabinets, 51

Microorganisms,
  classifications according to risk, 45 *et seq.*
  for experiments in schools, 47
  infections with, 43 *et seq.*
  supply for teaching, 26, 55
Microbiology,
  teaching, special hazards in, 54
  training and supervision, 54
Misuse of Drugs Act, 6
Mollusca, 32
Mutagens, 80 *et seq.*

Nettles, stinging, 37
Noise (Control of Pollution) Act, 6, 7

Occupational Exposure Limits, 79
Occupiers' Liability Act, 8
Opportunists, 43

Packaging and Labelling Regulations, 6
Parasites, supply for teaching, 26
Pathogenic microorganisms, 43, 61
Pesticides, 63 *et seq.*
Petroleum (Consolidation) Act, 6
Phytophotodermatitis, 37
Plant Pests (Great Britain) Order, 8
Plants,
  hazardous, 37 *et seq.*
  precautions against, 40
  Import and Export Order, 7
  infection from, 40
  irritant, 37 *et seq.*
  physical damage from, 40
  treatment of, 41
Poison ivy, 38
  oak, 38
Poisonous Substances in Agriculture Regulations, 64
Poisons,
  in agriculture, 63 *et seq.*
  protection and first aid, 64–65
  regulations, 5

Poisons Act, 6
  List Order, 6
  Rules, 6
Pollution, Acts and Regulations, 6
Polychlorinated biphenyls, 6
Portuguese man-o-war, 34
Power
  cables, 60
  take-off shafts, 62
Primula rash, 38
Protective clothing,
  in agriculture, 64
  in animal houses, 16, 20
  in microbiology, 53
Prohibition Notices, 4
Public Health Acts, 6, 7
  (Drainage of Trade Premises) Act, 6
Pupils, experiments involving, 29

Rabies (Importation of Dogs etc.)
  Order, 19
Radioactive Substances Act, 5, 6
Reptiles, hazardous, 32
Risk,
  areas and zones, 70, 71
  assessment,
    in biological laboratories, 68
    with microorganisms, 44
Rivers (Prevention of Pollution) Act, 6

Safety,
  Committees, 3
  Representatives, 3
    and Safety Committee Regulations, 3
Self-propelled equipment, 63
Short Term Limit Values, 78
Skin, infection through, 44
Snakebite, 35
Solvents,
  comparative properties, 82
  flammable, 77
  storage, 71
Spiders, 34
Spraying,
  in agriculture, 60, 63
  in glasshouses, 65
Storage,
  chemicals, 84
  gas cylinders, 71, 72
  solvents, 71
Suspect carcinogens, 79

Teratogens, 80 *et seq.*
Tetanus, 19
Threshold Limit Values, 78
Thunderstorms, 59
Toads, 35
Toxic chemicals, 78 *et seq.*
Tractors, hazards from, 61–63

Wasps, 34
Waste disposal,
  Acts and Regulations, 6
Wind, 60

Vapours, explosive and flammable, 77
Venomous animals, 3
  symptoms and treatment of injuries, 33
Vertebrates, capable of injuring man, 17, 32
Vicarious liability, 8

Waste material, disposal of,
  Acts and Regulations, 6
  agricultural, 65
  animal house, 20, 23
  chemical, 85
  contractors, 86
  microbiological, 52
Water Resources Act, 6

Zones, risk areas, 70, 71
Zoonoses Order, 7